Lecture Notes in Electrical Engineering

Volume 76

Gerard Cybulski

Ambulatory Impedance Cardiography

The Systems and their Applications

 Springer

Dr. Gerard Cybulski
Department of Applied Physiology
Medical Research Centre of Polish Academy of Sciences
5 Pawinskiego Str.
02106 Warsaw
Poland
e-mail: gerard@cmdik.pan.pl

and

Department of Mechatronics
Institute of Metrology and Biomedical Engineering
Warsaw University of Technology
8 Św. A. Boboli Str.
02525 Warsaw
Poland
e-mail: g.cybulski@mchtr.pw.edu.pl

ISSN 1876-1100

e-ISSN 1876-1119

ISBN 978-3-642-26659-1

ISBN 978-3-642-11987-3 (eBook)

DOI 10.1007/978-3-642-11987-3

Springer Heidelberg Dordrecht London New York

Cover design: eStudio Calamar, Berlin/Figueres

Printed on acid-free paper

Springer is part of Springer Science+Business Media (www.springer.com)

To my wife EWA and only daughter OLA

Preface

This book gives the reader the necessary physical background of impedance cardiography and currently available systems for hemodynamics holter monitoring. It compares ambulatory impedance cardiography and other clinically accepted methods through an updated state-of-the-art. Besides describing the techniques, it also shows the clinical applications of them. The book is intended for graduate and postgraduate students, researches and practitioners interested in impedance cardiography and/or in ambulatory monitoring of vital signals in human. The monograph summarizes my research on the subject I have been developing for the last 15 years. It is based in a large portion on my habilitation thesis, published in *Polish Journal of Medical Physics and Engineering* (PJMPE, 2005). Hereby, I would like to thank to the Publisher authorities of PJMPE for their kind agreement to use the material from the thesis in this book.

When writing about the acknowledgments I would like to thank to all my tutors and mentors I found on my scientific way. Without their help, inspiration and constant encouragement it would be very hard to make science. I have in mind many names, but let me mention only few of them: Prof. Krystyna Nazar, who looked after my physiological researches and promoted my Ph.D. thesis, Prof. Tadeusz Pałko, who inspired me to use impedance cardiography in physiology, Prof. Ryszard Grucza, the protector of my first scientific steps. On that occasion my special thanks are addressed to Dr. Wiktor Niewiadomski, my every day consultant and first critical reviewer of the scientific ideas. Without his help and long, interesting discussions some ideas would never appear.

I also would like to pass many thanks to the several Polish cardiologist of the younger generation associated with "Klub 30" of the Polish Cardiac Society. I have in mind co-authors of my papers, who inspired me and organised the clinical examinations essential to verify my impedance cardiography ambulatory monitoring system in clinical applications. I am grateful to Prof. Piotr Kułakowski, Dr. Edward Koźluk, Dr. Sebastian Stec, Prof. Przemysław Guzik, Dr. Olga Kruszelnicka, Dr. Ewa Michalak, Dr. Beata Zaborska and some other medical doctors.

Last but not least, I wish to thank my "lab-mates"—Anna Gąsiorowska, Ph.D., Anna Strasz, M.Sc., and Dorota Laskowska, M.Sc., who supported me despite the special way of my work.

Contents

List of Abbreviations

AC	alternating current
AF	Atrial fibrillation
AVd	atrio-ventricular delay
CO	cardiac output
CRT	cardiac resynchronisation therapy
DC	direct current
$(dz/dt)_{max}$	maximal amplitude of the first derivative impedance signal
ECG	electrocardiogram
EEG	electroencephalogram
ET (LVET, VET)	ejection time, left ventricular ejection time
EPCI	Ejection Phase Contractility Index.
EMS	Electro-mechanical systole
GSR	galvanic skin resistance
H	height of the subject in (cm)
HR	heart rate
LVH, RVH	left ventricular hypertrophy, right ventricular hypertrophy
LVET (ET, VET)	left ventricular ejection time
L_0	distance between receiving electrodes (cm)
PAT	paroxysmal atrial tachycardia
PCG	phonocardiogram
PEP	Pre-ejection period
ρ	resistivity
SV	stroke volume
VET (LVET, ET)	left ventricular ejection time
VEB	Ventricular extrasystole beat
ΔV	changes of the blood volume of the body segment (cm^3)
VEPT	Volume of Electrically Participating Tissues (a function of patient's gender, height and weight)

ΔZ	changes of the impedance of the segment limited by receiving electrodes (Ω)
Z_0	the base impedance of the segment limited by the receiving electrodes (Ω)

Chapter 1
Introduction

In this chapter is described the importance of the transient changes noninvasive monitoring of physiological variables in living objects. It also contains the rationale for the development of an ambulatory monitoring equipment including long-term recording of a heart mechanical activity.

1.1 The Importance of Monitoring Transient Changes

Living objects are able to adjust to changes in their environment and to demands generated by an organisms' own activity. This ability is the effect of complex, interacting control processes, which are manifested on the molecular, cellular, and systemic levels. Impairments of the processes of regulation are the causes or the results of many systemic diseases.

One of the methods of assessment of regulatory mechanisms' efficiency is analysis of the traces that characterise a particular systems' response to various stimuli during transient phases of the change between two steady states. There are two possibilities: the reaction is generated as an effect of a provocation applied to the organism (e.g. exercise, or a tilt test in humans) or the transient phase occurs during normal activity as a result of an unknown cause which clinicians aim to identify (e.g. cardiac arrhythmia, ischemia, etc.).

The effectiveness of the control of the flow of blood is a crucial condition of the correct development of physiological processes and the ability of the organism to undertake exercise. Blood flow may be controlled by two effectors: the heart and vascular system. Both are under control of an autonomic system, humoral influences and local autoregulatory processes.

There are several tests of the cardiovascular system in humans, among them orthostatic manoeuvre, tilt test, handgrip, dynamic exercise, Valsalva manoeuvre, and cold pressor test. The application of each test enables evaluation of

G. Cybulski, *Ambulatory Impedance Cardiography*,
Lecture Notes in Electrical Engineering, 76, DOI: 10.1007/978-3-642-11987-3_1,
© Springer-Verlag Berlin Heidelberg 2011

dysfunction of a particular system or impairment of adaptation abilities (orthostatic intolerance, autonomic system disorder, exercise capacity level, etc.). An advanced method for this type of diagnosis performed in humans is ambulatory monitoring of physiological parameters obtained during the subject's habitual activity.

1.2 Non-invasive Recording of the Cardiac Parameters and its Significance

The heart rate (HR), observed initially from a pulse, was the first parameter to be identified as describing the condition of the heart. HR detected from an electrocardiogram (ECG) is the most important vital signal in intensive care units. ECG also describes the state of the heart muscle and is the major source of information regarding the propagation of electric stimulation and neural control of cardiac activity.

The stroke volume (SV) is one of the important physiological parameters not fully available to clinicians (due to the invasive procedures necessary) or underestimated (when non-invasive methods are used). This parameter and its changes in response to physiological or pharmacological stimuli are potentially an excellent tool in evaluation of the mechanical efficiency of the heart. A reliable measurement of SV and its mean, e.g. in atrial fibrillation or in the presence of extrasystole, could be an indicator of effective cardiac work.

Cardiac output (CO), as a product of HR and SV, and changes in it are the physiological parameters describing the heart's efficiency and ability to undertake a load. It is also essential for assessing the progress of disease and, in e.g. heart failure, the monitoring of treatment. The analysis of hypertension cannot be considered seriously without accurate estimation of both blood pressure and CO.

Systolic time intervals (STI), mainly left ventricular ejection time (ET, LVET) and pre-ejection period (PEP) are the parameters describing the heart contractility. Numerous studies have shown that patients with myocardial disease (low ejection fraction, or other measures of left ventricle performance) had abnormal STI [1, 2, 3, 5, 8, 9]. This suggests that when invasive procedures are not available (typical dp/dt analysis) the STI measurement may be used, partially, as the indices of contractility.

Continuous blood pressure measurement [6] is now available also in an ambulatory version (Portapres). This signal is essential for the comprehensive analysis of the cardiovascular system e.g. in hypertension or baroreflex sensitivity evaluation.

1.3 Ambulatory Monitoring and Implementations of it

Ambulatory monitoring of electrical heart activity (ECG) has been used for more than 40 years, since Holter constructed his device enabling long-term recording of ECG signal on magnetic tape [4]. Ambulatory monitoring of ECG is a well-established technique applied in ambulatory and clinical practice to evaluate cardiac arrhythmia, myocardial ischemia, heart rate variability (HRV) and QT dispersion. In particular, this method makes it possible to detect transient, work and stressor dependent abnormalities in the cardiac activity of patients with correct resting ECG. Moreover, this method is useful in evaluation of applied therapy or during rehabilitation, as well as in checking of pacemakers.

These features have remained unchanged although the recording medium has improved from magnetic tape to solid-state chips and PC Flash Cards. Moreover, the number of signals recorded has increased. The idea of long term recording of the biological signals has been extended to ambulatory blood pressure monitoring (ABP or Portapres), to monitoring of signals of electrical activity in the brain (AEEG), to recording of skeletal muscle activity (AEMG), and to breathing and pulse oximetry (Oxyholter).

The most widely applied ambulatory technique is ECG monitoring. Despite the unquestioned usefulness of this method there are several circumstances in which 24-h monitoring of ECG does not supply sufficient information for evaluation of the heart's functioning. Simultaneous recording of ECG and of a signal that reflects central hemodynamics activity might solve this problem. It appears that electrical Impedance Cardiography (ICG), a simple method that allows for continuous, non-invasive determination of SV, maximum velocity of ejection, and ejection time (ET), could be used to supply such a signal.

1.4 Ambulatory Monitoring Using Impedance Cardiography Signals

The development of ambulatory monitoring methods is an implementation of the idea of measurement of vital biological signals during the normal daytime activity of patients when all stresses and natural conditions are preserved rather than during traditional clinical examinations. Ambulatory monitoring enables non-invasive, comprehensive recording of data and allows the evaluation of short-term transient events that are difficult to detect using any of the stationary methods.

Simultaneous recording over a long time of ECG and of a signal that reflects central hemodynamic activity may bring some further useful diagnostic data, particularly in arrhythmia patients and in pharmacological studies. Even with all the limitations of the technique it appears that electrical ICG, as a simple method allowing for continuous non-invasive determination of SV, estimation of maximum velocity of ejection $(dz/dt)_{max}$ and measurement of ET may be used to provide such a signal. Cardiac contractility and SV indices may be also applied for determination of hemodynamic efficiency in healthy individuals (in Sports Medicine or Exercise Physiology) or in patients during regular pacing and during arrhythmia or ischemia events for supplementary diagnosis or evaluation of pharmacological therapy.

The idea of ICG ambulatory monitoring with signal recording on memory chips was introduced in 1985 and 1987 by Webster's group, [7, 10]. Their solution allowed, however, only for short time recordings. Using more advanced PC Cards technology I have aimed to develop a holter-type system and verify its utility. I hope that my studies will help to promote inclusion of the ICG in the family of ambulatory methods. In my opinion, there is an almost unlimited number of possible fields of clinical and physiological applications for such a system. It could, for example, be used for monitoring the hemodynamic consequences of transient events, e.g. paroxysmal arrhythmia, for monitoring the hemodynamic efficiency of pacemaker stimulation and quantitative verification of changes in hemodynamic parameters caused by pharmaceutical stimuli. Such a system could be used to record transient events that are difficult or even impossible to visualize using other clinical methods, particularly during a patient's normal daily activity.

The described assumptions were the main reasons why in several research laboratories worldwide the independent groups decided to undertake the effort of creating the new ambulatory, portable, miniaturize system for hemodynamic signals analysis. In pursuing the main idea they decided to:

- develop a miniaturized holter-type ICG device for ambulatory monitoring of cardiac hemodynamics,
- develop analytical software which allows for automatic determination of cardiac parameters,
- verify the new systems by comparing the ambulatory measurements of cardiac parameters with data simultaneously recorded by clinically accepted non-invasive reference methods,
- evaluate the rate of artefacts during ambulatory ICG monitoring,
- present the possible clinical and physiological applications of the new systems.

References

1. Arshad, W., Duncan, A.M., Francis, D.P., O'Sullivan, C.A., Gibson, D.G., Henein, M.Y.: Systole-diastole mismatch in hypertrophic cardiomyopathy is caused by stress induced left ventricular outflow tract obstruction. Am. Heart J. **148**(5), 903–909 (2004)
2. Cybulski, G.: Dynamic impedance cardiography—the system and its applications. Pol. J. Med. Phys. Eng. **11**(3), 127–209 (2005)
3. Frey, M.A.B., Doerr, B.M.: Correlation between ejection times measured from the carotid pulse contour and the impedance cardiogram. Aviat. Space Environ. Med. **54**(10), 894–897 (1983)
4. Holter, N.J.: New method for heart studies. Science. **134**, 1214–1220 (1961)
5. Lewis, R.P., Boudoulas, H., Leier, C.V., Unverferth, D.V., Weissler, A.M.: Usefulness of the systolic time intervals in cardiovascular clinical cardiology. Trans. Am. Clin. Climatol. Assoc. **93**, 108–120 (1981)
6. Penaz, J., Voigt, A., Teichmann, W.: Contribution to the continuous indirect blood pressure measurement (in German). Gesamte Inn. Med. **31**(24), 1030–1033 (1976)
7. Qu, M., Webster, J.G., Tompkins, W.J., Voss, S., Bogenhagen, B., Nagel, F.: Portable impedance cardiograph for ambulatory subjects. In: Proceedings of the Ninth Annual

Conference of the IEEE Engineering in Medicine and Biology Society, IEEE. pp. 1488–1489, vol. 3. New York (1987)
8. Weissler, A.M., Peeler, R.G., Roehll, W.H.: Relationship between left ventricular ejection time, stroke volume, and heart rate in normal individuals and patients with cardiovascular disease. Am. Heart J. **62**, 367-78 (1961)
9. Weissler, A.M., Harris, W.S., Schoenfeld, C.D.: Systolic time intervals in heart failure in man. Circulation **36**(2), 149–159 (1968)
10. Zhang, Y., Qu, M., Webster, J.G., Tompkins, W.J.: Impedance cardiography for ambulatory subjects. In: Proceedings of the Seventh Annual Conference of the IEEE/Engineering in Medicine and Biology Society. Frontiers of Engineering and Computing in Health Care, IEEE, pp.764–769, vol. 2. New York (1985)

Chapter 2
Impedance Cardiography

In this chapter the necessary physical background of impedance cardiography will be presented, especially regarding the electrical properties of the biological tissues, the methods applied for measurement of the bioimpedance and the inaccuracy the models and limitations of the technique.

2.1 Bioimpedance Measurement: Applications and Importance

The biological tissues are the complex, anisotropic conductors with a resistive and reactive components. Researches are interested in both, the value of the impedance and their changes, since it could be the source of the useful diagnostic information. The value of the impedance depends on the type of the analysed tissue whereas the changes, especially cyclic changes, could be caused by the organs (or tissues) translocation, shape and structure modifications, and/or the volume and location of intracellular fluid. The value and the changes of the impedance of the biological tissues depends also on the frequency of applied current.

Currently there are three major fields of medical applications of the measurement of electrical properties of biological tissues:

- differentiation between the tissues (impedance spectroscopy),
- recognition of the pathological processes in the tissue basing on the impedance versus frequency relations (impedance spectroscopy).
- analysis of the function of the organs (e.g. flow evaluation in heart, brain, extremities, etc.),
- visualization of the internal body structure (using multielectrode electroimpedance tomography—EIT).

G. Cybulski, *Ambulatory Impedance Cardiography*,
Lecture Notes in Electrical Engineering, 76, DOI: 10.1007/978-3-642-11987-3_2,
© Springer-Verlag Berlin Heidelberg 2011

Impedance cardiography is a diagnostic method based on measurement of the electrical properties of the biological tissues applied to the thorax region. In that monograph I will focus only on the application of the impedance method to measure the hemodynamic activity of human heart when ambulatory (portable, wearable, Holter-type) systems are used.

2.2 Electrical Properties of the Biological Tissues

The electrical properties of the biological tissues are depend on the presence of body fluids. Moreover, those properties could be different for in vivo and in vitro analysis of the same tissue. The solid part of the tissue is mainly based on cellular membranes, which have the isolating properties. The presence of conductive fluid part and solid, isolating, component in the same tissue is affecting on anisotropic, electrical properties of the living tissue. Figure 2.1 (based on the results published by [67] presents the idealized curve of the changes in relative permittivity (ε_r) of biological tissue with applied frequency. There are three major frequency ranges describing the changes of relative permittivity called the dispersion α, β and γ. In some tissues not every range is clearly represented. Dispersion α, occurring in lower frequencies (around 100 Hz) is associated with interfacial polarization and ion transport across the cell membranes. The β-dispersion is related to the "capacitive shorting-out of membrane resistances, and rotational relaxations of biomacromolecules". The γ-dispersion is origins from "the relaxation of bulk water in the tissue" [67]. The electrical properties of the tissues is analysed using impedance spectroscopy method. It allows to differentiate between the various tissues and the level of degeneration caused by the pathological processes, e.g. tumour development, ischemia and the process of rejection the transplanted organ [64, 95]. The mainstream researches on bioimpedance methods are presented regularly during the International Conferences on Electrical Bio-Impedance (ICEBI) organised worldwide every third year since the first meeting in 1969. The basic research achievements in that field was presented in the monograph written by Grimnes and Martinsen [28].

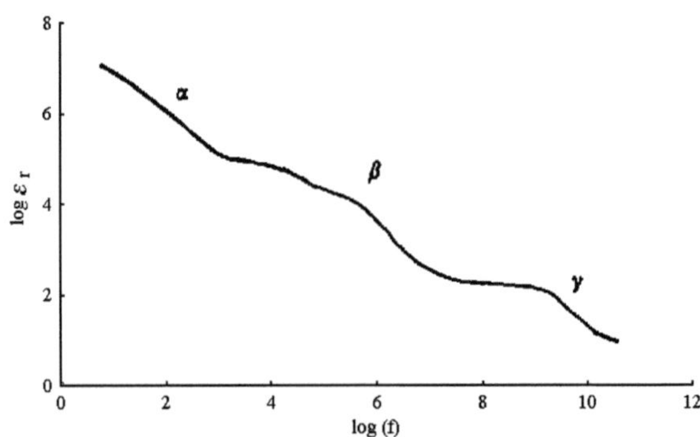

Fig. 2.1 An idealized curve of the changes in relative permittivity of biological tissue with applied frequency

Table 2.1

Type of the tissue	Resistivity (Ω cm)
Blood plasma	63
Blood (for hematocrit Ht $= 47\%$)	150
Skeletal muscle (longitudinal)	300
Skeletal muscle (transverse)	700
Heart muscle (dog)	750
Lungs (dog)	1,200
Fat	2,180
Saline 0.9%	57

In the impedance cardiography it is used the frequency range between 20 and 100 kHz. For that range the relative permittivity is at the level of several thousands (e.g. $\varepsilon_r = 5{,}000$ for $f = 50$ kHz). The thin cellular membranes separating the conductive fluids create the electrical capacitances at the level up to 1 μF/cm^2. This results in a resistive and capacitive character of the electrical properties of the biological tissues, with the absence of the symptoms of the inductive components. The lowest resistivity of biological material is observed in blood plasma (63 Ω cm). However considering the tissues blood has the lowest resistivity (150 Ω cm for Ht $= 47\%$) and the highest one is observed for fat (1,275 Ω cm). The blood resistivity play a major role in the measurement of cardiac performance using impedance methods. It was assumed that the blood resistivity has an isotropic properties in static conditions, however there are some suggestions that flowing blood is characterized by a relative decreasing of resistivity by up to 10%. Moreover, some authors basing on experimental studies [96] and also model analysis claims that the anisotropic properties of flowing blood are much pronounced that we initially assumed and suggested to use the resistivity tensor.

Table 2.1 presents the resistivity of different biological tissues and fluids basing on Geddes and Baker [23] compendium. The main conclusion drawn from the resistivity data presented in Table 2.1 is that in the impedance cardiography the change of the impedance signal is generated mainly by the changes caused by blood volume translocations. Other tissues are either not changing the volume or have at least 2 times higher resistivity. In the case of the heart muscle it is even 5 times higher, which leads to the conclusion that heart "isolates" the blood accumulated inside it.

2.3 Tissue as a Conductor

The tissues could be treated as a finite, inhomogeneous volume conductor [44]. For the impedance cardiography applications we can also assume that it is a sourceless medium. The tissue impedance could be represented by three element model with

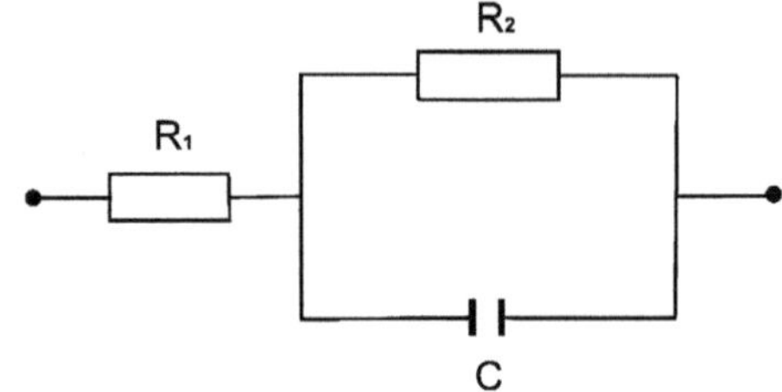

Fig. 2.2 The tissue impedance represented by three element model with single time constant

single time constant (Fig. 2.2). For this model it could be proposed the function, according to Cole–Cole model [13]. This function shows impedance Z as a function of frequency with real (R) and imaginary components (jX): $Z(f) = R + jX$. In the expanded form it is presented below:

$$Z(f) = R_\infty + \frac{R_0 - R_\infty}{1 + j\varpi\tau} = R_\infty + \frac{R_0 - R_\infty}{1 + \varpi^2\tau^2} - j\varpi\tau\frac{R_0 - R_\infty}{1 + \varpi^2\tau^2}$$

where $Z(f)$ impedance as a function of frequency, R_0 resistance at the frequency $f = 0$ Hz, R_∞ resistance at the frequency $f = \infty$, τ time constant (RC_2).

The current passing through the biological tissues is described by the continuity condition

$$div(\sigma\,gradj) = 0,$$

where σ, spatial conductivity of the tissue, φ, electrical potential

The Geselowitz theorem [26, 27] proposed the method of calculation the changes in impedance caused by the changes in volume, conductivity and movement of the organs in the body. Lehr [42] proposed the vector derivation method. Geselowitz work was more generalised by Mortarelli [56] considering modified geometry of the objects.

2.4 Frequency and Current Values

There is general agreement to use the frequencies from the range 20-100 kHz and the sinusoidal current of the amplitude between 1 and 5 mA [92]. However some designers used a lower than suggested values of current. The lower boundary application is the suggestion leading to obtain the sufficient signal-to-noise ratio. The 1 mA current can create the muscle excitation at the lower than 20 kHz frequency. Also the skin–electrode impedance at 100 kHz is 100 times lower than in low frequencies. This helps to diminish the unwanted impact of the changes in the skin-electrode impedance, occurring during motion, into measured cardiological signal. However increasing the frequency of the application current above the 100 kHz causes the problem of stray capacitances.

2.5 Bioimpedance Measurement Methods

2.5.1 Biopolar and Tetrapolar Method

There are two main methods of the bioimpedance measurement: bipolar and tetrapolar. In the bipolar method there are two electrodes which play a role of application and receiving in one. Near the electrodes the current density is higher than in other parts of the tissue, which results in a non-uniform impact into the total impedance measurement. The total impedance signal is a superposition of two components: the skin–electrode impedance (modified by blood flow-induced movement) and the original signal (e.g. caused by the blood flow). In practice it is difficult to separate them. The scheme of the bipolar impedance measurement is presented in Fig. 2.3.

In a four-electrode (tetrapolar) the application electrodes and receiving are separated. Figure 2.4 presents the scheme of the tetrapolar impedance measurement, typical way of obtaining the impedance cardiography signal. The constant amplitude current oscillates between the application electrodes (A) and the voltage changes are detected on the receiving electrodes (R). This voltage, due to the constant amplitude of current is proportional to the impedance of the tissue segment limited by the band electrodes. The voltage changes, are proportional to the impedance changes between receiving electrodes. The main advantage of that method over the bipolar is that the current density distribution is more uniform. Another one is that the disturbing role of the electrode impedances is minimized.

2.5.2 Alternating Constant-Current Source

The measured impedance and its changes is in series with stray impedances of the electrodes and cables. Webster [92] estimates the stray capacitances at 15 pF which results in impedance as large as 100 k (at the frequency 100 kHz of the oscillating current. Since the measured value has the impedance smaller than 30–35 (and its

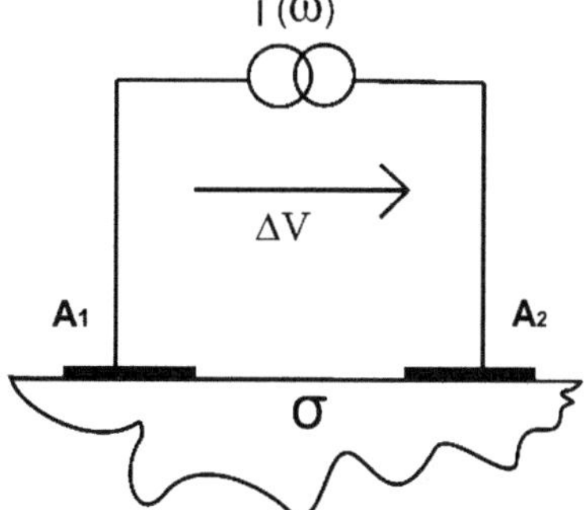

Fig. 2.3 The scheme of the bipolar impedance measurement. The constant amplitude current oscillates between the application electrodes and the voltage changes are detected on the same electrodes A1 and A2

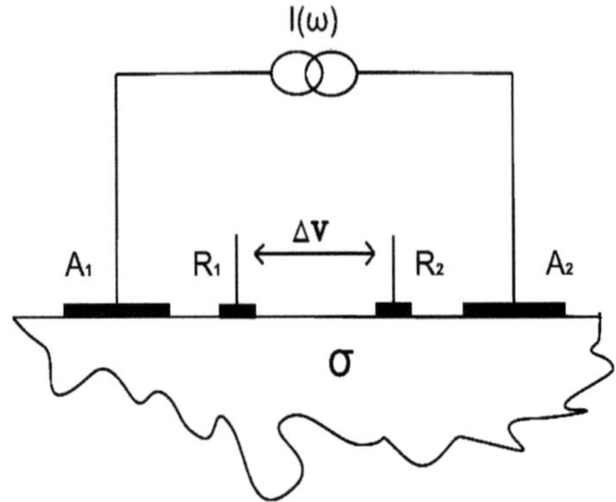

Fig. 2.4 The scheme of the tetrapolar (four-electrode) impedance measurement. The constant amplitude current oscillates between the application electrodes (A1–A2) and the voltage changes are detected on the receiving electrodes (R1–R2)

changes are about 100 times lower) so the internal impedance of the alternating current source should be sufficiently high. In some applications the constant amplitude oscillating current is supplied via transformer with a low-capacity.

2.5.3 Receiving Unit

The voltage is sensed at the receiving electrodes. Assuming that their stray impedances are at the same level as the application electrodes (100 k at the frequency 100 kHz), the input impedance of the detecting amplifier should be sufficiently high. The instrumentation amplifiers offered by many suppliers could be used to design the first stage of the circuit that satisfy this requirements. Assuming the current amplitude $i = 1$ mA and the base impedance $Z = 20$–35 Ω, we can assume that the maximal value of the voltage amplitude at the level $v = Zi$ (20–35 mV). The most important signal is stroke volume calculations is the change in the main impedance. The observed change of the impedance (ΔZ) has an amplitude of 100–400 mΩ, which results in 0.1–0.4 mV signal for the 1 mA current. Increasing the amplitude of the applied current to 5 mA we could bring the base impedance signal (Z) and the change of the impedance (ΔZ) to the level of 100–175 and 0.5–2 mV, respectively. However, increasing the current amplitude is possible in the stationary system but very much limited in battery powered applications like Holter-type instrumentation [97].

2.5.4 Demodulation Unit

The useful, small biological signal of low frequency is hidden in the large signal of the high frequency (20–100 kHz). Thus any amplitude demodulator AM could be used to extract the voltage signal proportional to the changes in the in the impedance. Some designers uses the phase-sensitive detectors which are insensitive to the mains frequency. At the output of the demodulator occurs the signal proportional to the sum of Z and ΔZ signals. So another functional unit is essential to provide separated Z and ΔZ signals. The Z base impedance (denoted later by Z_0) is proportional to the content of blood in the segment limited by the receiving electrodes, whereas ΔZ is used to estimate stroke volume, systolic time intervals and derivative signals.

2.5.5 Automatic Balance Systems

Any impedance system is very sensitive to the motion artefacts. The changes in the skin–electrode impedances caused by the body movement may result in the changes larger then physiologically justified variations. This may caused the saturation of the amplifier which could last even for more than 10 s. In stationary solutions the operator could manually initiate the adjustment unit which allows the momentary resets the ΔZ to zero. In the ambulatory solutions this process should be performed automatically, just to avoid the long time loss of the signal. The automatic balance system which distinguish the physiologically induced changes from the unwanted variations (caused by e.g. motion artefacts) for stationary solutions was described by Shankar and Webster [76]. The comparator-based solution was applied in the ambulatory system [14].

2.6 Electrodes Types and Topography

There are several types of the electrodes topography used in generating of the impedance cardiography signals. Historically, the first was a set of four band electrodes. However, then occurred some modification of that techniques.

2.6.1 Band Electrodes, Spot Electrodes and Mixed Spot/Band Electrodes

When band electrodes are applied, two of them were placed around the neck separated by few centimetres and one over the suprasternal notch and the last one 10 cm lower. Spot electrodes are placed at the same level on the chest as the band electrodes same positions as band electrodes. Figure 2.5 presents the typical position of band electrodes connected to the main parts of each impedance cardiography system—generator and receiver (amplifier/detector). Please note the distance between inner-receiving electrodes L_0. The ECG electrodes are not presented. Some systems use the same electrodes for detecting ECG and generating ICG signal. Several researchers use the mixed topology of the electrodes substituting band electrodes by the easy to use, however more noisy spot electrodes (Fig. 2.6).

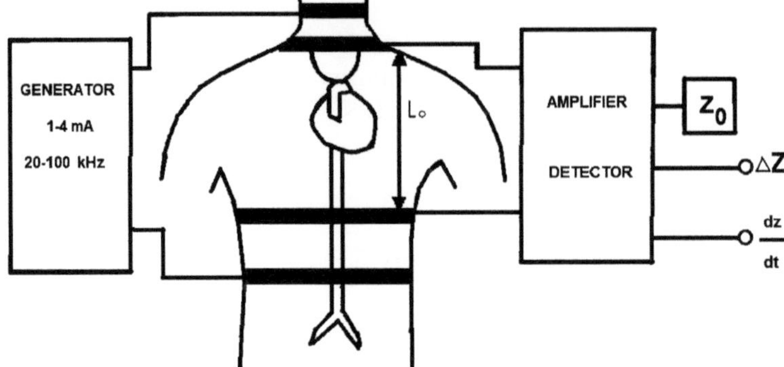

Fig. 2.5 The typical position of band connected to the main parts of each impedance cardiography system—generator and receiver (amplifier/detector) according to the scheme proposed by Kubicek et al. [39]. Please note the distance between inner-receiving electrodes L_0. The ECG electrodes are not presented

Fig. 2.6 The topography of the electrodes used in the PhysioFlow Enduro device

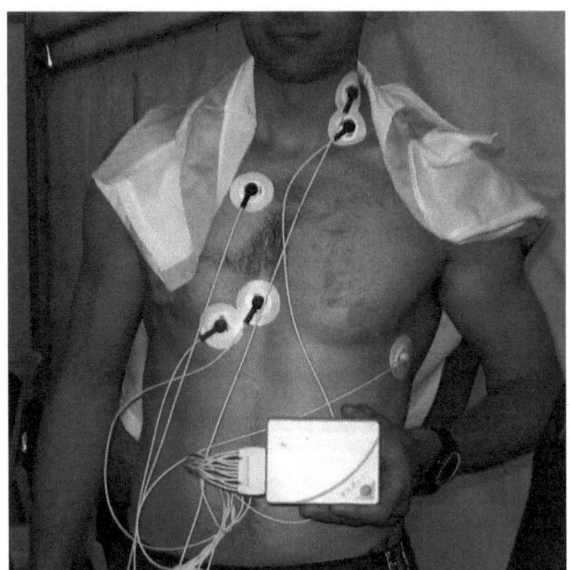

2.6.2 Other Solutions

The BioZ (and derivative) systems producers, as well as Task Force Monitor and PhysioFlow suggested their own topography of the electrodes. They could be found on the web pages of the respective producers (http://www.sonosite. com/products/bioz-dx, http://www.cnsystems.at/product-line/task-force-monitor/ features). Below please find the figure adapted from the materials on Physio-Flow Enduro device published on their web page.

2.7 Signal Description and Analysis

2.7.1 Impedance Cardiography Traces

The typical impedance cardiography traces accompanied by the one lead of ECG are presented in Fig. 2.7. On the first channel ECG is shown, on the second the first derivative of the ΔZ signal which is denoted dz/dt. Calibrations for the respective signals are also presented. In practice only ECG and the first derivative signal (dz/dt) are used to calculate the hemodynamic parameters. Another data essential to make those evaluations is the value of the base impedance (Z_0), which is usually not changing very fast. So in some applications Z_0 was stored as a only one value for several cycles.

Figure 2.8 shows the traces of electrical (ECG-top line) and mechanical activity of the heart (dz/dt—bottom line) recorded for the one cycle of the heart. There are some characteristic points (waves) defined and the method of determination for some important periods of the heart cycle presented. All of those symbols and time periods determination are described in the following sub-chapters.

2.7.2 Characteristic Points on Impedance Cardiography Curves

Following the method of description the characteristic points and waves in ECG traces using letters (PQRST) researchers suggested the similar way for impedance signal (dz/dt). Lababidi et al. [41] associated the some notches in the ICG (dz/dt curve) with the events in mechanical action of cardiac muscle. They

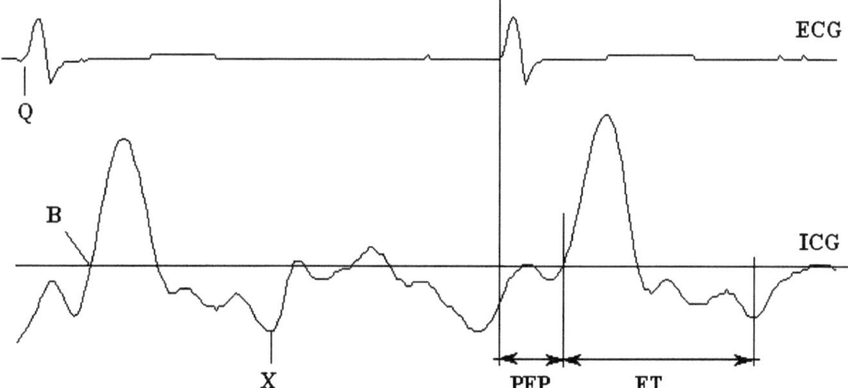

Fig. 2.7 The typical impedance cardiography traces: the changes in the dz/dt (first derivative of the ΔZ) signal denoted as dz/dt (2nd channel), recorded simultaneously with one lead of ECG (1st channel). Please note the way of determination of PEP, ET (LVET) and $(dz/dt)_{max}$, here marked as "amp" (adapted from [16])

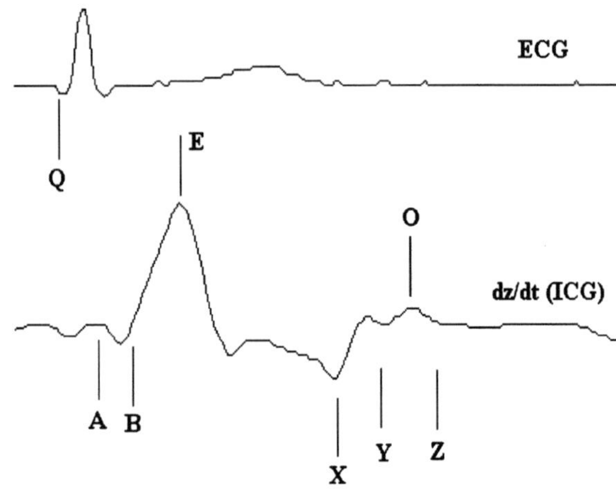

Fig. 2.8 The traces showing electrical (ECG-*top line*) and mechanical activity of the heart (dz/dt-*bottom line*) recorded for the one cycle of the heart. *Symbols* denote some characteristic points on the ECG and impedance traces and the definition of periods occurring during heart cycle

suggested A, B, $(\mathrm{d}z/\mathrm{d}t)_{\max}$, (sometimes denoted as E), X, Y, O and Z. Although there are some trials in drawing the conclusions from the qualitative differences in the morphology of the ICG signal for various groups of cardiac patients (as for the ECG signal) they still remains in the stage of "attempts" and ICG is mainly used for quantitative estimation of cardiac parameters. Those "attempts" are mainly associated with the diagnosis of the valves malfunction.

Below are listed and described the main waves and notches of ICG signal.

A (atrial wave)   This wave is associated with the atrial contraction and its amplitude is correlated with the ejection fraction of the left atrium. The modification in time coupling between P-ECG and A-ICG is especially pronounced in patients with heart blocs [79]. A-wave is not observed when atrial fibrillation occurs [88].

B   This wave is associated with the opening of the aortic valve. It occurs on the ascending part of the ICG curve before $(\mathrm{d}z/\mathrm{d}t)_{\max}$. It is assumed that B wave occurs at the level of baseline for the ICG curve. There are some suggestions that its length is correlated with the isovolumetric contraction time. Denotes the beginning of the ejection time (left ventricular ejection time). The precise identification of a B point on the impedance cardiograph waveform is an important for accurate calculation of systolic time intervals, stroke volume, and cardiac output [45, 47, 49, 50]. Sometimes identifying the location of the B point is problematic because the characteristic upstroke that serves as a marker of this point is not always apparent [43]. To avoid this problem it was introduced a reliable method for B point identification, based on the consistent relationship between the R to B interval and the interval between the R-wave and the peak of the dZ/dt function (called by them as RZ) [43]. Thy suggested the function relating RB to RZ (RB = 1.233RZ −

$0.0032RZ^2 - 31.59$) which accounts for 90–95% of the variance in the B point location across ages and gender and across baseline and stress conditions. As they claim, this relation affords a rapid approximation to B point measurement that, in noisy or degraded signals, is superior to visual B point identification and to a derivative-based estimate [43].

E $((dz/dt)_{max})$: This point is placed on the top position of the curve reflecting the maximal speed of the change in the impedance. It is associated with the maximal velocity of the ejection measured by using ultrasound methods [33]. There is sometimes a collision between the meaning of the $(dz/dt)_{max}$ as a top point on the ICG curve and $(dz/dt)_{max}$ considered as a maximal amplitude of the ICG signal measured from the baseline to the top of the curve. Perhaps the top point should be denoted only by E letter. Some other authors call them Z wave or just dz/dt_{max}

X   Corresponds to the closure of the aortic valve (and second heart sound—S2). Described as a minimal value of dz/dt signal, occurring after maximum E. In some patients it is not well pronounced (older people) or is placed on the same level as Y point showing the bimodal pattern of this wave. This might cause to the misleading automatic determination of the ejection time

Y   Corresponds to the closure of pulmonic valve

O   This wave is associated with changing of the volume during the diastolic phase of the cycle and opening snap of the mitral valve. The high value of the O wave (similar to E) might be the symptom of the bicuspid valve malfunction or the heart inefficiency during heart acute ischemia [80]

Z   This wave describes a decreeing following the O wave. It is associated with a third heart sound—S3

2.7.3 Characteristic Periods in Impedance Cardiography

Q-B   Time between the onset of the electrical contraction of the chambers and the opening of the aortic valve. It is described as pre-ejection period (PEP) which is a sum of electromechanical delay (EMD) and isovolumetric contraction (IVC)

R-E (R-Z)   Time between the R point in ECG and the maximal value of the impedance signal. Seigel et al. [75] found the high correlation coefficient ($r = 0.88$) between R-E and the time from R-ECG to maximal value in the first derivative of the pressure signal (dp/dt_{max})

	for the left ventricle. It might be treated as an index of the left chamber performance
Q-E	Time between the Q in ECG and the maximal value of the dz/dt signal. It used in calculation of so called Heather index of cardiac contractility [29]
B-X	Time between the opening and closing of the aortic valve described as left ventricular ejection time or shortly (ejection time)—LVET (ET)

The most frequently used are PEP and ET since they have the direct association with cardiac systole and were earlier analysed using the polyphysiographic methods [36, 78].

2.7.4 Hemodynamic Indices

There are some indices which could be calculated solely using ICG and ECG traces and in the other group there are indices calculated using the above mentioned and other parameters like characterising mean arterial blood pressure (MAP) and body surface area (BSA). There were introduced and/or described by many researchers [39, 41, 68, 85, 94]

SV	Stoke volume (SV) expressed in millilitres [ml] is the basic volumetric parameter characterising the amount of blood ejected from the left ventricle in one cycle
CO	Cardiac output (expressed in [l min^{-1}] or [dm^3 min^{-1}]) is the volume of blood pumped by the heart during 1 min. It is a product result of SV and heart rate (HR). Of course it could be also calculated as a sum of all SV values occurring during 1 min period
STR	Systolic time ratio (PEP/ET), the ratio of the two basic systolic time intervals considered as an index of cardiac contractility
TFC	Total fluid content is the index of the presence of fluid in the thorax region. Practically it is the inverse value of the basic impedance Z_0 expressed in the units of 1/kΩ or [(kΩ)$^{-1}$]
Heather index	The cardiac contractility index defined as a ratio of the maximal amplitude of dz/dt signal to the Q-E period (expressed in [Ω/s^2]. So it means the rate of impedance changes to the time in which they are performed from the electrical onset to the peak mechanical action

The different producers provide some other indices using the mentioned above characteristic points and periods evaluated from the ICG curve. Other hemodynamic indices are created as a normalised of those from the basic group or an arithmetical combination of them and other hemodynamic parameters. The important normalisation factor is body surface area (BSA) expressed in [m^2] units and calculated using the Dubois and Dubois [19] formula:

$$\text{BSA} = 0.007184 \cdot W^{0.425} \cdot H^{0.725},$$

where W is weight supplied in [kg], H is height in [cm].

Stroke index (SI)	SV can be indexed to a patient's body size by dividing by body surface area (BSA)
Cardiac index (CI)	CO indexed to a patient's body size by dividing by body surface area (BSA)
Systemic vascular resistance (SVR)	can be calculated if cardiac output (CO), mean arterial pressure (MAP), and central venous pressure (CVP). SVR = (MAP–CVP)/CO. Since CVP is close to 0, it could be simplified to MAP/CO. The units are [(dyne × s)/cm^5] or [MPa s/m^3]. Typical values of SVR: 900–1,200 dyn s/cm^5 (90–120 MPa s/m^3). Some authors use name of total systemic resistance (TSR)
Systemic vascular resistance index (SVRI)	is an SVR normalized by division of SVR by BSA
Velocity index (VI) and acceleration index (ACI)	There are two specific parameters introduced by producers of the BioZ impedance system. VI is the maximum rate of impedance change, and is representative of aortic blood velocity. ACI is the maximum rate of change of blood velocity and representative of aortic blood acceleration
Left cardiac work (LCW)	An indicator of the amount of work the left ventricle must perform to pump blood each minute. This index reflects myocardial oxygen consumption. It is the product of blood pressure and blood flow. LCW is determined with the following equation: LCW = (MAP–PAOP) × CO × 0.0144, where PAOP is Pulmonary Artery Occluded Pressure, or wedge pressure. When Pulmonary Artery Catheter is not applied, a default value of 12 mm Hg can be used because of PAOP's minimal effect on LCW determination. LCW is measured in Watts [W] or other units of work
Left cardiac work index (LCWI)	LCW can be indexed to a patient's body size by substituting Cardiac Index for Cardiac Output to yield LCWI = (MAP–PAOP) × CI × 0.0144
Stroke work index (SWI)	pressure times blood volume ejected in one beat normalised by body surface area (BSA)

Mean systolic ejection rate (MSER)	It is a volume of blood ejected during ejection time (divided by LVET) and normalised by body surface area (BSA)
Vascular rigidity (VR)	The ratio of pulse pressure per stroke volume index (SVI)
Ejection phase contractility index (EPCI)	It is defined as $EPCI = (dz/dt)'_{max} TFC$. Sramek suggested that the aortic blood peak flow is a mirror image of ejection phase myocardial contractility. Thus the rate of cardiovascular impedance changes over time (dz/dt) is an image of the aortic blood flow

Thus, there are different types of the indices based on the ECG and ICG signals regarding cardiac timing, contractility and another group of the indices constructed basing an the ICG signals and blood pressure and/or body characterizing parameters. I am sure that researchers will develop some other indices showing their correlation with physiological variables or finding the new exponents of cardiac work, contractility or the efficiency of the heart pump.

2.7.5 The Influence of Breathing

Breathing modulates the impedance signal causing its fluctuation around the base impedance Z_0 which results in "moving baseline" of dz/dt signal. This might result in the misleading absolute value calculations of SV when the "stiff zero level" is used to find the maximal amplitude of the dz/dt trace. Thus the base line should be determined separately for each cycle or this modulating effect of breathing should be removed from the signal.

Slow fluctuations of the ICG signal around the base line caused by breathing could be eliminated by using the following methods:

1. temporary elimination of breathing, used by Dennistone et al. [17], Keim et al. [31] and Du Quesnay et al. [18];
2. application of an ensemble averaging method to provide the "mean cycle" for the certain period [48, 58];
3. digital filtering [51, 91].

Du Quesnay et al. [18] noted that temporary elimination of breathing affects SV despite the reduction of breathing modulation. Moreover, it is not possible to perform some physiological tests when not breathing (e.g. exercise tests). Application of an ensemble averaging technique means that information regarding beat-to-beat changes is lost. Thus filtering is the best method of reducing the interference of breathing modulation with the ICG signal.

2.7.6 The Origin of the Impedance Cardiography Signals

In impedance cardiography it is usual to use alternating current of frequency in the range of $f = 20$–150 kHz and with constant amplitude (range 0.5–5 mA), oscillating between application electrodes A1–A2 (Fig. 2.5) and meeting the impedance of a significant real component (15–30 Ω) and a negligible (causing a $10°$ phase shift) imaginary component.

Blood is a tissue characterised by the highest conductivity. Its resistivity (130–160 Ω cm) is two times lower than the resistivity of muscle tissue (300 Ω cm), which is second on the list ranged from low to high values. Also, blood's resistivity is many times higher than the resistivity of other tissues [23, 24]. This biophysical property of the blood means that the impedance between receiving electrodes is mainly caused by the volume of blood contained within this segment of the chest.

Nyboer's formula (explained below) was a mathematical description of a simple model of the changes in blood vessels in the chest during cardiac cycle. It presented the chest, as a uniform cylinder in which there was a single blood vessel of a particular diameter filled with blood. The blood ejection from heart to the aorta during contraction was simulated as a rapid extension of this vessel (uniform increase of diameter) and the similarly, rapid uniform return after closing the aortic valve. The time of the vessel extension was equal to the ejection time. This simplified model of the heart hemodynamics phenomena and its consequences when SV was calculated were many times criticised by both enthusiasts, [7] and Porter and Swain [69], and opponents of the ICG method, [30] and Keim et al. [31]. This criticism was an inspiration for studies where the impact of particular components into ICG signal was evaluated [2, 9, 34, 38, 65, 74, 77, 90].

The validity of one-cylinder model was questioned by [71], who suggested the serial connected two-cylinder model instead. They studied the length (distance between the receiving electrodes) dependence of the impedance parameters Z_0, $(dz/dt)_{max}$, and stroke volume (SV) in both models. It was shown that, within a one-cylinder model, all parameters are directly proportional to the length, whereas, if the volume conduction of the thorax and the neck are modelled separately, Z_0 and $(dZ/dt)_{min}$ were expected to be linear dependent and SV was non-linear upon the length. Those expectations were compared to results from in vivo measurements [71] using an electrode arrays placed o the thorax. The results showed a nearly linear relation between the ICG parameters and the length except for small distances. Regression analysis of the linear part revealed statistically significant intercepts ($p < 0.05$). When stroke volume was calculated using Kubicek's equation, neither the intercept nor the non-linear part could be explained by a one-cylinder model, whereas a model consisting of two cylinders serially connected described the experimental results accurately. Thus they concluded that SV estimation based on a one-cylinder model is biased due to the invalid one-cylinder model and suggested the corrections for the Kubicek equation using their two-cylinder model [71].

It may be concluded that the following mechanical phenomena occurring during cardiac cycle and give impact to the impedance cardiography signal when the alternating current passes the chest:

1. the aorta's and the neck artery's extension caused by arterial pulse pressure,
2. the pulmonary vessels' extension,
3. changes in the volume of heart and the volume of blood filling it,
4. changes in the volume of blood filling the pulmonary vessels, resulting in an increase of the lungs' conductivity,
5. changes in the blood resistivity in the large vessels caused by reorientation of the blood cells as a function of the velocity of blood flow,
6. changes in skeletal muscle resistivity caused by the pulse blood flow.

Shankar et al. [77], Kosicki et al. [38] and Patterson [65], presented a mathematical model of the impedance changes in the chest that occur during the cardiac cycle. Kim et al. [34], based on the finite element model and analysing the impact of the factors described in points 1, 3, 4, 5. They concluded that changes in impedance are proportional to the extension of the aorta and that modification of the lung's resistivity has an eight times' smaller impact than that of the aorta. The impact of the heart signal on base impedance was significant but not proportional to changes in heart volume. These findings were confirmed by Kosicki et al. [38], who presented a cylindrical model of the chest with non-coaxial positioning of the chest organs. They noticed the small phase of the aortic signal in comparison to changes of resistivity during blood flow. They also presented the magnitude and phase shift of the particular components of the impedance signal. Patterson [65], however, on the basis of a three-dimensional resistors model concluded that the impact on the impedance signal of particular components is: the pulmonary component (61%), main arteries (23%), skeletal muscle (13%), and other sources (3%). Most of these papers were theoretical and checked only on "phantom" physical models. The problem of whether the impact of the aortic or the pulmonary artery is higher was solved by Thomsen [90], who experimentally found a higher correlation between impedance SV and aortic blood pressure ($r = 0.63$) than between SV and pulmonary artery blood pressure ($r = 0.26$). Additionally experiments performed by Bonjer et al. [9], on dogs (isolating the heart and lungs in a rubber bag) showed that heart muscle has a negligible impact to the ICG signal.

There were several theoretical and experimental studies performed with the aim to determine the origin of the impedance signal. Penney [66], basing on several studies, summarised the contribution of the size and function of anatomical structures into the impedance cardiography signals. The result of this resume is presented in Table 2.2.

Mohapatra [53] after the analysis of several hypotheses regarding the origin of the impedance cardiography signals concluded that the impedance changes were caused only by cardiac hemodynamics. He suggested that the signal reflects both a change in the blood velocity and change in blood volume. Moreover, the changing speed of ejection has its major effect on the systolic part of ΔZ whereas the

Table 2.2 Origin of the impedance signal in impedance cardiography (based on [66]

Contributing organ	Contribution (%)
Pulmonary artery and lungs	+60
Aorta and thoracic musculature	+60
Right ventricle	−30
Left ventricle	−30
Pulmonary vein and left atrium	+20
Vena cava and right atrium	+20

changing volume (mainly of the atria and great veins) affects the diastolic portion of the impedance curve. Amidst the controversies regarding the unclear source of impedance cardiography signal, pointed out again by Mohapatra [54] and some other researchers, the correlation coefficients between the values obtained by ICG and reference methods has acceptable levels.

On the basis of the results of these papers several conclusions could be drawn:

1. The ICG signal is complex and its components are not synchronised in phase and direction,
2. The proportions between its components are not well defined,
3. The ICG signal is dependent on the changes in the diameter of the aorta, neck arteries and pulmonary vessels and in the resistivity of flowing blood caused by blood cell reorientation (only in the large vessels),
4. The ICG signal appears not to be dependant on heart volume and the amount of blood in it (isolating properties of the heart muscle in relation to blood resistivity),
5. For typically positioned electrodes the ICG signal is not dependent on pulmonary artery extension.

2.7.7 The Methods of Stroke Volume Calculation

There are different methods of calculation a stroke volume (and in the consequence other hemodynamic parameters) using impedance signal, the characteristic points on the impedance waveform and parameters describing the physical dimension of the analysed segment of the human body. The usage of the different formulas may lead to the marked scattering of the results. Historically, the first was Kubicek formula [39, 40] derived from the Nyboer works [62, 63] signal. Let me describe shortly some of the formulas and methods using the "historical" order.

Nyboer Formula

Atzler and Lehman [3], for the first time suggested that changes in electrical impedance of the chest are related to the blood volume translocation in the thorax observed during the cardiac cycle. Their investigations were developed by Nyboer et al. [62] and Nyboer [63], who presented a formula describing the relationship between changes in blood volume in any segment of the body and the changes in its impedance:

$$\Delta V = \rho \cdot L_0^2 \cdot Z_0^{-2} \cdot \Delta Z$$

where ΔV, changes of the blood volume of the body segment [cm^3]; ρ, blood resistivity [Ω cm]; L_0, distance between receiving electrodes [cm]; Z_0, basic impedance of the body segment limited by receiving electrodes [Ω]; ΔZ, changes of the impedance of the segment limited by receiving electrodes [Ω].

It is generally accepted that changes in the thoracic impedance ΔZ are caused mainly by the ejection of blood from the left chamber to the aorta and are proportionate to the stroke volume (SV). Kubicek et al. [39] suggested this interpretation in 1966.

Kubicek Formula

Kubicek suggested modification of Nyboer's formula (2.1) replacing $\Delta Z = (dz/dt)_{max} \cdot ET$, and substituting $\Delta V = SV$, in a cardiac version of the impedance method [39, 40]. This resulted in establishing the basic impedance cardiography formula named after Kubicek:

$$SV = \rho \cdot L_0^2 \cdot Z_0^{-2} \cdot (dz/dt)_{max} \cdot ET$$

where SV stroke volume [cm^3], $(dz/dt)_{max}$, the maximum of the first derivative of the impedance signal [Ω/s], ET, ejection time [s], time of blood ejection from the left chamber, determined by selection of characteristic points on (dz/dt) trace, (other symbols are explained with Nyboer formula above).

Sramek Formula

Sramek proposed another method of calculating stroke volume using 3 components: volume of electrically participating tissues (VEPT—which is a function of patient's gender, height and weight), ventricular ejection time (VET), which has the similar meaning as LVET or ET in Kubicek formula and the ejection phase contractility index (EPCI), which is a product of maximal amplitude of the dz/dt signal $(dz/dt)_{max}$ and TFC (which is an Z_0^{-1}). His idea was to show in the formula that SV is directly proportional to the physical size of a patient (i.e., to VEPT—body habitus scaling constant), directly proportional to duration of delivery of blood into the aorta (i.e., to VET), and (SV is directly proportional to the peak aortic blood flow (i.e., to EPCI). This lead to the formula:

$$SV = VEPT \times VET \times EPCI$$

where VEPT, volume of electrically participating tissues (a function of patient's gender, height and weight); VET, ventricular ejection; EPCI, ejection phase contractility index.

When substituting the above symbols by the expression used in the Kubicek formula it gives:

$$SV = \frac{(0.17H)^3}{4.25 \cdot Z_0} \cdot (dz/dt)_{max} \cdot ET$$

where H, the height of the subject in [cm], $(dz/dt)_{max}$, is the maximal amplitude of the dz/dt signal [Ω/s], ET, left ventricular ejection time [s], Z_0, the base impedance of the segment limited by the receiving electrodes [Ω].

Sometimes this formula is presented as:

$$SV = \frac{L_0^3}{4.25 \cdot Z_0} \cdot (dz/dt)_{max} \cdot ET$$

where L_0, the distance between receiving electrodes [cm]; L_0, is assumed to be 0.17 of the height of the subject. This resulted in occurrence of the factor $0.17H$ in the earlier formula.

Sramek-Bernstein Formula

Some researchers use another formula for SV calculation called the Sramek-Bernstein equation:

$$SV = \delta \cdot \frac{(0.17H)^3}{4.25 \cdot Z_0} \cdot (dz/dt)_{max} \cdot ET$$

It is based on the assumption that the thorax is a truncated cone with length L and circumference C measured at the xiphoid level [4, 5, 84, 86]. It was checked that C/L ratio is equal to 3.0 regardless of age or sex (with the exception of newborns). Also L is assumed to be 0.17 of the height (H) and a correction factor (δ) relating actual and ideal weight was introduced.

Bernstein and Lemmens [6], suggested another formula called N (Bernstein). The formula is given below:

$$SV = \frac{V_{ITBV}}{\xi^2} \cdot \sqrt{\frac{(dz/dt)_{max}}{Z_0}} \cdot ET,$$

where V_{ITBV}, 16 $W^{1.02}$ [ml], empirical formula for intra-thoracic blood volume estimation when W is expressed in [kg]. Dimensionless index of transthoracic aberrant conduction; the way of its determination is given in the mentioned paper [6].

They verified the results using N (Bernstein) with those obtained by thermodilution method in 106 cardiac postoperative patients and achieved the better accordance between the measurements in comparison to the usage of the other formulas.

The Kubicek, Sramek and Sramek-Bernstein equations are based on different methodological assumptions but both are able to provide a reliable SV estimation.

TaskForce Monitor Method

The Kubicek formula is the consequence of the cylindrical model of the thorax applied in the theoretical considerations. Sramek noted that this model is too simple to give the precise determination of the SV. He abandoned the cylindrical model of thorax and assumed that the thorax is the frustum of a parameters dependant on some anthropometric parameters. Than introduced the volume of electrically participating tissues (VEPT), which is a function of patient's gender, height and weight. The task force monitor, producers of stationary equipment, followed the way of the usage of anthropometric measures to estimate the electrically participating volume of the thorax named by them as (V_{th}) [22].

They noted that "shape of the body is neither an exact cylinder nor a frustum but more or less determined by the fact of whether the patient is underweight, normal or obese: underweight people will tend to have a more cylindrical thorax shape while obese people will have a more frustum-shaped thorax". Than they suggested that "the grade of leanness/obeseness can be estimated by the body mass index (BMI), whereby a BMI of 25 is considered to be the border between normal and marginal overweight" [22].

They used a tilt tests to determine the influence of the body composition as well as the base impedance Z_0 on V_{th}. Thus the V_{th} is described according to the formula:

$$V_{th} = C_1 \cdot H^3 \cdot \frac{BMI^n}{Z_0^m},$$

where C_1, powers m and n are subject to proprietary non-disclosure [22].

Since BMI is defined as $W/H^{2,}$ the following equation was implemented:

$$SV = C_1 \cdot H^3 \cdot \frac{\left(\frac{W}{H^2}\right)^n}{Z_0^m} \frac{(dz/dt)_{max} \cdot ET}{Z_0},$$

and after simplifications:

$$SV = C_1 \cdot \frac{W^n \cdot H^{3-2n}}{Z_0^{m+1}} \cdot (dz/dt)_{max} \cdot ET$$

PhysioFlow Method

The PhysioFlow designers did not reveal the exact formula for SV calculations. The idea of that calculations were presented in the Appendix I of the paper by Charloux et al. [11].

In contrary to the previous formulas they did not use the baseline values of the impedance Z_0 or the physical parameter like a distance between the receiving electrodes L_0, although they use BSA (so weight and height of the subjects). They used, however, the Haycock formula,

$$\mathrm{BSA} = 0.024265 \; W^{0.5378} H^{0.3964}$$

for BSA instead of that provided by Dubois and Dubois [19]. There are several steps in calculation of SV (or SVI) using PhysioFlow (they called SVI by SVi). Basing on the mentioned Appendix 1 from [11] let me describe those steps. A first evaluation of SVi, called $\mathrm{SVi_{cal}}$, is computed during a calibration procedure based on 24 consecutive heart beats recorded in the resting condition. The largest impedance variation during systole (Z_{max}–Z_{min}), and the largest rate of variation of the impedance signal $(\mathrm{d}Z/\mathrm{d}t)_{max}$, called the contractility index (CTI) is stored. The SVi calculation depends on the ventricular ejection time (ET). The designers of the PhysioFlow have chosen to use a slightly different parameter, called the thoracic flow inversion time (TFIT), expressed in [ms]. The TFIT is the time interval between the first zero value following the beginning of the cardiac cycle (starting from QRS in ECG) and the first nadir after the peak of the ejection velocity ($\mathrm{d}Z/\mathrm{d}_{max}$). Afterwards, the TFIT is weighted [W(TFIT)] using a specific algorithm where the pulse pressure (PP) (the difference between systolic arterial pressure and diastolic arterial pressure) and the momentary HR is used. They introduced their method of SV calculations basing on the assumptions that the aortic compliance contributes to the signal waveform. For example, Chemia et al. [12] have demonstrated the existence of a linear relationship between aortic compliance and the SV/PP ratio. In the algorithm the PP, calculated from a sphygmomanometer measurement, is introduced at the end of the Physio Flow calibration phase. Similarly, certain in relationship with. Second foundation of their method is the of the influence of the oscillatory and resonance phenomena on the ICG signal morphology. They spotted that Murgo et al. [57] described a relationship between the pressure waveform and aortic impedance or momentary HR. Thus they used the HR as a second factor entering into the algorithm.

As a result of the above concepts, $\mathrm{SVi_{cal}}$ is computed according to the following formula:

$$\mathrm{SVi_{cal}} = k\left[(\mathrm{d}Z/\mathrm{d}t_{max})/(Z_{max} - Z_{min})\right]W(\mathrm{TFIT_{cal}})$$

where k is a constant, and the subscript "cal" indicates the parameters measured during the calibration phase. Thus $\mathrm{SVi_{cal}}$ represents the baseline reference. During the data acquisition phase, the variations of the parameters described above are analysed and compared to those obtained during the calibration procedure. For instance, the designers demonstrated that the SV variations result mainly from a combination of contractility fluctuations (CTI) or $(\mathrm{d}z/\mathrm{d}t)_{max}$ changes and of TFIT variations. So, the stroke volume index is calculated basing on the calibrated value of SVi and the factor based on TFIC, CTI and their calibrated values:

$$\mathrm{SVi} = \mathrm{SVi_{cal}} \left((\mathrm{CTI/CTI_{cal}}) \cdot (\mathrm{TFIT_{cal}/TFIT}) \right)^{1/3}$$

They claim that this concept is supported by a study by Moon et al. [55], who showed that changes in SV, for example during exercise, are correlated with variations in dz/dt, but inversely correlated with variations in left ventricular ejection time. They noted that in all equations used by other impedance cardiography devices, these two parameters appear as a product. The main advantage of that formula over other is that positioning of the electrodes is not critical, since Z_0 evaluation is unnecessary.

2.7.8 Blood Resistivity Impact

Since blood resistivity (ρ) is a proportional factor in SV calculations that use the Kubicek formula its unbiased determination is important for evaluation of SV. There are two approaches to this problem in the literature: assumption that r is constant and in the range of 130–150 Ω cm, [20, 40, 58, 69, 81], or determination of r as a second order or exponential function of hematocrit (Hct), [17, 24, 37, 89].

The supporters of the first approach have used resistivity closer to 130 Ω cm. Quail et al. used a transformed Kubicek formula to determine blood resistivity in dogs for different Hct and measured SV using a magnetic flowmeter [70]. For $\pm 35\%$ of Hct changes (range 0.22–0.66, mean 0.41) they observed only $\pm 3.3\%$ change in resistivity (range 141.3–132.2 Ω cm, mean 136.8 Ω cm). Moreover, the blood resistivity as a function of Hct was determined in vitro. However, Swanson and Webster estimated that blood resistivity is positively correlated with blood velocity and may change by 10% within the physiological range of the flow [87]. Thus it seemed that application of a constant value of r gives a smaller error than the introduction of another measurement (resistivity), which is out of control. Wtorek et al. also pointed out the anisotropy of the resistivity of flowing blood [95, 96].

The ICG method is sensitive to artefacts, so motion, anxiety, restlessness, shivering, and hyperventilation may interfere with measurements and modify physiologic responses. Factors preventing good electrode-to-skin contact (sweating, oils, and severe obesity) may also limit the accuracy of signal detection.

2.8 Signal Conditioning

When it is not possible to eliminate the source of the noise the most important method for improving the signal to noise ration in impedance cardiography measurements is the averaging technique.

2.8.1 Ensemble Averaging Method

In impedance cardiography the ensemble averaging is a method for reduction of the influence of the artefacts or unwanted effect of other signals. It is used as a method of improvement a signal to noise ratio, when "noise" means not only the typical disturbances in signal detection (e.g. caused by motion) but also the signals modulating the original source of physiological changes. The crucial issue in that process is the method of synchronisation. When there is no other signal (like in ECG quality improvement) the gating is performed by the characteristic wave in the transformed signal (R-ECG). The respective wave in first derivative of the impedance signal (dz/dt) would be the peak of it called dz/dt_{max} (some authors use dz/dt_{min} notation, just to mark that the direction of impedance changes is negative although usually presented as a positive). However, the peak value of dz/dt is also affected by shape modifications (including bimodality). Thus the signal averaging is usually synchronized by the signal of electrical activity of the heart—by R-ECG. It is performed over a certain period (e.g. 60 s) by summing the digitised samples gated by R-wave peak and dividing by the numbers of beats in the analysed period. Another approach of synchronizing by the peak value of the first derivative signal (dz/dt) was applied by Kim et al. [35]. They tested their method during four stages treadmill exercise. Stroke volumes of the five subjects averaged by the peak dz/dt were 0–23.5% higher than those obtained by the R point at rest and exercise. The ensemble averaging is used to reduce the influence of natural beat-to-beat variability and the disturbing effect of the respiration on impedance signal. The ensemble averaging was used in the stationary ICG by some authors [48, 58] and it is used also ambulatory version of ICG in VU-AMS [72]. This method may be used in on-line analysis or off-line calculations. Muzzi et al. [58] founded in stationary systems that ensemble-averaging of transthoracic impedance data provides waveforms from which "reliable estimates of cardiac output can be made during normal respiration in healthy human subjects at rest and exercise and in critically ill patients". Kelsey and Guethlein [32] compared cardiac parameters determined from ensemble averaged signals to those determined by simple beat-to-beat averaging over 60 and 20-s sampling intervals. They confirmed "validity of ensemble averaging as a method for deriving impedance cardiographic measures of myocardial performance". In one of the first research systems for portable ICG monitoring the signal averaging was performed on-line and the result was stored in the memory without the original trace, due to the memory size limitation [93]. The loss of the original signal was the main disadvantage of that method. Nowadays this restrictions disappeared due to the common availability of the memory cards in a variety of formats. Just to underline the tremendous progress in this area it is worth to mention that in 1993 the PCMCIA 20 MB Card produced by SanDisk (maximum capacity available on the market) cost about $1500. The cost of ultrafast compact flash card of 4 GB capacity in 2010 is below $50, and is decreasing. The largest capacity of 64 GB is sold at the price of below $350. Also the data transfer is very fast, currently at the rate of 20–30 MB/s. So the ensemble averaging method is used only to improve the signal to noise ratio and extract the trends in long-term hemodynamic monitoring but not because of the memory capacity limitations.

2.8.2 Large-Scale Ensemble Averaging Method

The large scale ensemble averaging method is a way of averaging longer than over the 60 s period. It was proposed by the group who developed the VU-AMS device [72]. This method was applied as a off-line analysis for a period of the similar type of the patient activity but lasting not longer than 1 h. If the activity type lasted longer than 1 h (sleeping) it was divided into several periods shorter than 1 h. The periods were selected using the entries from a patient diaries describing the activity, physical load, posture, location (home, work, etc.), and social situation. The full description of the categories is given by Riese et al. [72]. The morphology analysis of the large scale ensemble averaging was performed in a similar way to a 60 s averaging.

The reason why the ensemble averaging method has been introduced was the demand of a precise, artefact free determination of Pre-ejection Period (PEP). PEP is a short systolic time interval beginning at Q-in QRS wave of ECG and ending at the moment of aortic opening derived from any signal showing the mechanical heart activity. The location of Q wave is not creating the significant error. But in case of ICG signal the ambiguous reading of the aortic valve opening might be a reason of substantial inaccuracy of PEP interval determination. When we assume roughly the range of PEP at the level 50–150 ms, the inaccuracy of 5 ms in the determination of that period creates a maximal error of 10% in one direction. However, 5 ms is not a maximal value of the inaccuracy in PEP determination. Ensemble averaging and/or large scale of the ensemble averaging are the methods leading to the PEP measurement improvement, minimizing the inaccuracy. Why accurate PEP measurement is so important? There are some suggestions that PEP absolute level correlates with sympathetic drive [10, 21]. Thus PEP would be an easy measure of sympathetic activity, useful in psychophysiological reactions to the stressors. Moreover, it was found that prolonged PEP occurs in patient with aortic stenosis [52]. Since PEP is strongly positively related to the level of venous return (pre-load) it could be also considered as a measure of it [82]. Also PEP expressed as a fraction of left ventricular ejection time (LVET) could be treated as a measure of cardiac contractility [1, 45].

2.9 Technical Aspects of ICG-Limitations, Errors and Patients' Safety

Evaluation of SV by ICG, like any indirect measurement method, is biased by the superposition of partial errors. Assuming that errors of determination of L_0 and Z_0

are 1%, for ET -1.7% (5 ms of the time resolution for 200 Hz sampling frequency), and for $(\mathrm{d}z/\mathrm{d}t)_{\max} - 2\%$ (resistivity is set constant) the minimal SV error could be estimated by a differential method to be at the level of about 4%. Swanson and Webster, noted that it is not possible to identify all sources of errors but gave technical requirements for instrumentation which limited the measurement error to 5% [87]. They suggested that the frequency of the application current should be within the range $f = 20...150$ kHz, the current amplitude $i = 0.5...5$ mA, the common mode rejection ratio of the input amplifier—CMRR >400, the input impedance—$R_{\mathrm{in}} > 4$ kΩ, the output impedance of the current generator $R_{\mathrm{out}} > 20$ kΩ, and the noise level $V_{\mathrm{n}} < 0.5$ µV. The range of current is limited by the acceptable level of noise and the current density $(j \leq 5 \mathrm{~mA~cm}^{-2})$. The frequency range is limited by the bioelectric properties of the tissues.

Apart from errors due to indirect measurement and the technical limitations of the method other sources of errors are associated with the methodology employed—the simplified model of the chest, the disturbing influence of breathing and uncertainties in estimation of blood resistivity.

2.10 Modifications of ICG, and Other Impedance Techniques

The specific modifications of the impedance technique are rheo-angiography and rheo-encephalography [64]. Another application of the impedance technique is an impedance multi-frequency spectroscopy used to characterise tissue [64]. Also some researchers are involved in development of electro-impedance tomography (EIT), including the electrical mammography [61].

2.11 Physiological and Clinical Applications of Impedance Cardiography

The ICG method has been applied to evaluation of changes in cardiac output during exercise in many different studies, starting from those reported by Miyamoto et al. [48, 50, 51], Miles et al. [46]. Bogaard et al. [8] published a review of the possibilities of hemodynamic measurements by ICG during exercise. They concluded that, although ICG "derived stroke volume calculation is based on several debated assumptions, numerous validation studies have shown good accuracy and reproducibility, also during exercise". Moreover, [73], in their review stated that "impedance cardiography is becoming an accepted method for safe, reliable, and reproducible assessment of hemodynamics in heart failure".

ICG has been successfully used in many studies performed in the author's home laboratory on cardiovascular response to the handgrip [25], orthostatic manoeuvre [15] and other physiological tests including dynamic exercise [16, 98].

Also, the effects of 3 days of bed rest on hemodynamic responses to submaximal loads during graded exercise in athletes and men with sedentary lifestyles was analysed [83].

The controversies around the verification of ICG resulted in a sceptical approach by health authorities to this technique, which used to be considered a research and not clinical method. Thus Medicare and Medicaid Services and health insurance companies in USA used not to reimburse the cost of ICG usage [99]. This approach was changed (from 1 July 1999) and revised US policy on cardiac output monitoring by electrical impedance now allows limited coverage of cardiac monitoring using electrical bioimpedance, a form of plethysmography, for six uses:

- Suspected cardiovascular disease
- Fluid management
- Differentiation of cardiogenic from pulmonary causes of acute dyspnoea
- Optimisation of pacemaker's atrioventricular interval
- Determination of need for IV inotropic therapy
- Post-transplant myocardial biopsy patients

Insurance contractors may cover additional uses when they believe there is sufficient evidence of the medical effectiveness of such uses.

Not covered is the use of such a device for any monitoring of patients with proven or suspected disease involving severe regurgitation of the aorta, or for patients with minute ventilation (MV) sensor function pacemakers, since the device may adversely affect the functioning of that type of pacemaker. Moreover these devices do not render accurate measurements in cardiac bypass patients when they are on a cardiopulmonary bypass machine though they do provide accurate measurements prior to and post bypass pump. Their detailed description may be found at http://new.cms.hhs.gov/manuals/downloads/Pub06_PART_50. pdf.

Following the decision of Medicare and Mediaid Services in USA the respective national health institutions in other countries decided to allow refunding the impedance cardiography diagnosis, initially in a very limited areas, e.g. Intensive Care Units only. Certainly, these decisions are stimulating for the producers of the ICG equipment and hopefully will affect the development of the hemodynamic ambulatory monitoring systems based on impedance cardiography method.

2.12 Conclusions

All of the ways of stroke volume calculation were verified using the clinically accepted reference techniques. Despite the methodological differences, real and assumed level of the error for this indirect measurement and the used model, it seems that the absolute values of the stroke volume plays a minor role in ambulatory monitoring of the impedance signal. In that particular situation the patient is the reference for itself and the starting level of the SV is not so important when e.g. the hemodynamic effectives of the heart in arrhythmia events is visualised. In similar cases the percentage of the decrease in stroke volume could be give quantitative information on the acceptable level [59, 60]. In several clinical situations even a qualitative evaluation of the $(dz/dt)_{max}$ amplitude in neighbouring cycles (e.g. normal vs. trigeminy) would provide an additional diagnostic data.

References

1. Ahmed, S.S., Levinson, G.E., Schwartz, C.J., Ettinger, P.O.: Systolic time intervals as measures of the contractile state of the left ventricular myocardium in man. Circulation **46**, 559–571 (1972)
2. Anderson Jr, F.A., Penney, B.C., Patwardhan, N.A., Wheeler, H.B.: Impedance plethysmography: the origin of electrical impedance changes measured in the human calf. Med. Biol. Eng. Comput. **18**, 234–240 (1980)
3. Atzler, E., Lehmann, G.: Uber ein neues Verfahren sur Darstellung der Hertztatigkeit (Dielectrographi). Arbeitsphysiologie **5**, 636–639 (1932)
4. Bernstein, D.P.: A new stroke volume equation for thoracic electrical bioimpedance: theory and rationale. Crit. Care Med. **14**(10), 904–909 (1986)
5. Bernstein, D.P.: Continuous noninvasive real-time monitoring of stroke volume and cardiac output by thoracic electrical bioimpedance. Crit. Care Med. **14**(10), 898–901 (1986)
6. Bernstein, D.P., Lemmens, H.J.: Stroke volume equation for impedance cardiography. Med. Biol. Eng. Comput. **43**(4), 443–450 (2005)
7. Boer, P., Roos, J.C., Geyskes, G.G., Mees, E.J.D.: Measurement of cardiac output by impedance cardiography under various conditions. Am. J. Physiol. **237**(4), H491–H492 (1979)
8. Bogaard, H.J., Woltjer, H.H., Postmus, P.E., de Vries, P.M.: Assessment of the haemodynamic response to exercise by means of electrical impedance cardiography: method, validation and clinical applications. Physiol. Meas. **18**(2), 95–105 (1997)
9. Bonjer, F.H., van den Berg, J., Dirken, M.N.: The origin of the variations of body impedance occurring during the cardiac cycle. Circulation **6**(3), 415–420 (1952)
10. Cacioppo, J.T., Uchino, B.N., Berntson, G.G.: Individual differences in the autonomic origins of heart rate reactivity: the psychometrics of respiratory sinus arrhythmia and preejection period. Pychophysiology **31**(4), 412–419 (1994)
11. Charloux, A., Lonsdorfer-Wolf, E., Richard, R., Lampert, E., Oswald-Mammosser, M., Mettauer, B., Geny, B., Lonsdorfer, J.: A new impedance cardiograph device for the non-invasive evaluation of cardiac output at rest and during exercise: comparison with the "direct" Fick method. Eur. J. Appl. Physiol. **82**(4), 313–320 (2000)
12. Chemia, D., Hebert, J.L., Coirault, C., Zamani, K., Suard, I., Colin, P., Lecarpentier, Y.: Total arterial compliance estimated by stroke volume-to-aortic pulse pressure ratio in humans. Am. J. Physiol. **274**, H500–H505 (1998)
13. Cole, K.S., Cole, R.H.: Dispersion and absorption in dielectrics. I. Alternating current field. J. Chem. Phys. **1**, 341–351 (1941)
14. Cybulski, G., Książkiewicz, A., Łukasik, W., Niewiadomski, W., Pałko, T.: Ambulatory monitoring device for central hemodynamic and ECG signal recording on PCMCI flash memory cards, pp. 505–507. Computers in Cardiology, IEEE, New York, NY, USA (1995)
15. Cybulski, G.: Influence of age on the immediate cardiovascular response to the orthostatic manoeuvre. Eur. J. Appl. Physiol. **73**, 563–572 (1996)

16. Cybulski, G.: Dynamic Impedance cardiography—the system and its applications. Pol. J. Med. Phys. Eng. **11**(3), 127–209 (2005)
17. Dennistone, J.C., Maher, J.T., Reeves, J.T., Cruz, J.C., Cymerman, A., Grover, R.F.: Measurement of cardiac output by electrical impedance at rest and during exercise. J. Appl. Physiol. **30**, 653–656 (1976)
18. Du Quesnay, M.C., Stoute, G.J., Hughson, R.L.: Cardiac output in exercise by impedance cardiography during breath holding and normal breathing. J. Appl. Physiol. **62**(1), 101–107 (1987)
19. DuBois, D., DuBois, E.F.: A formula to estimate the approximate surface area if height and weight be known. Arch Intern. Med. **17**, 863–871 (1916)
20. Ebert, T.J., Eckberg, D.L., Vetrovec, G.M., Cowley, M.J.: Impedance cardiograms reliably estimate beat-by-beat changes of left ventricular stroke volume in humans. Cardiovasc. Res. **18**, 354–360 (1984)
21. Esler, M., Jennings, G., Lambert, G., Meredith, I., Horne, M., Eisenhofer, G.: Overflow of catecholamine neurotransmitters to the circulation: source, fate, and functions. Physiol. Rev. **70**(4), 963–985 (1990)
22. Fortin, J., Habenbacher, W., Heller, A., Hacker, A., Grüllenberger, R., Innerhofer, J., Passath, H., Ch, Wagner, Haitchi, G., Flotzinger, D., Pacher, R., Wach, P.: Non-invasive beat-to-beat cardiac output monitoring by an improved method of transthoracic bioimpedance measurement. Comput. Biol. Med. **36**(11), 1185–1203 (2006). (Epub 2005 Aug 29)
23. Geddes, L.A., Baker, L.E.: Specific resistance of biological material—a compendum of data for biomedical engineer and physiologist. Med. Biol. Eng. **5**(3), 271–293 (1967)
24. Geddes, L.A., Sadler, C.: The specific resistance of blood at body temperature. Med. Biol. Eng. **5**, 336–339 (1973)
25. Grucza, R., Kahn, J.F., Cybulski, G., Niewiadomski, W., Stupnicka, E., Nazar, K.: Cardiovascular and sympatho-adrenal responses to static handgrip performed with one and two hands. Eur. J. Appl. Physiol. Occup. Physiol. **59**(3), 184–188 (1989)
26. Geselowitz, D.B.: On bioelectric potentials in an inhomogeneous volume conductor. Biophys. J. **7**(1), 1–11 (1967)
27. Geselowitz, D.B.: An application of electrocardiographic lead theory to impedance plethysmography. IEEE Trans. Biomed. Eng. **18**(1), 38–41 (1971)
28. Grimnes, S., Martinsen, Ø.: Bioimpedance and Bioelectricity Bbasics, 2nd edn. Elsevier-Academic Press, Amsterdam-London. ISBN:978-0-12-374004 (2008)
29. Heather, L.W.: A Comparison of Cardiac Output Values by the Impedance Cardiograph and Dye Dilution Techniques in Cardiac Patients. Progress Report no. Na 594500. National Aeronautics and Space Administration, Manned Spacecraft Center, Houston (1969)
30. Ito, H., Yamakoshi, K.I., Togawa, T.: Transthoracic admittance plethysmograph for measuring cardiac output. J. Appl. Physiol. **40**(3), 451–454 (1976)
31. Keim, H.J., Wallace, J.M., Thurston, H., Case, D.B., Drayer, J.I., Laragh, J.H.: Impedance cardiography for determination of stroke index. J. Appl. Physiol. **41**(5 Pt. 1), 797–799 (1976)
32. Kelsey, R.M., Guethlein, W.: An evaluation of the ensemble averaged impedance cardiogram. Psychophysiology **27**(1), 24–33 (1990)
33. Kerkkamp, H.J.J., Heethaar, R.M.: A comparison of bioimpedance and echocardiography in measuring systolic heart function in cardiac patients. Ann. N Y. Acad. Sci. (Issue: Electrical Bioimpedance Methods: Applications to Medicine And Biotechnology) **873**, 149–154 (1999)
34. Kim, D.W., Baker, L.E., Pearce, J.A., Kim, W.K.: Origins of the impedance change in impedance cardiography by a three-dimensional finite element model. IEEE Trans. Biomed. Eng. **12**, 993–1000 (1988)
35. Kim, D.W., Song, C.G., Lee, M.H.: A new ensemble averaging technique in impedance cardiography for estimation of stroke volume during treadmill exercise. Front. Med. Biol. Eng. **4**(3), 179–188 (1992)
36. Kizakevich, P.N., Teague, S.M., Nissman, D.B., Jochem, W.J., Niclou, R., Sharma, M.K.: Comparative measures of systolic ejection during treadmill exercise by impedance cardiography and Doppler echocardiography. Biol. Psychol. **36**(1–2), 51–61 (1993)

37. Kobayashi, Y., Andoh, Y., Fujinami, T., Nakayama, K., Takada, K., Takeuchi, T., Okamoto, M.: Impedance cardiography for estimating cardiac output during submaximal and maximal work. J. Appl. Physiol. **45**(3), 459–462 (1978)
38. Kosicki, J., Chen, L., Hobbie, R., Patterson, R., Ackerman, E.: Contributions to the impedance cardiogram waveform. Ann. Biomed. Eng. **14**, 67–80 (1986)
39. Kubicek, W.G., Karnegis, J.N., Patterson, R.P., Witsoe, D.A., Mattson, R.H.: Development and evaluation of an impedance cardiac output system. Aerosp. Med. **37**(12), 1208–1212 (1966)
40. Kubicek, W.G., Patterson, R.P., Witsoe, D.A.: Impedance cardiography as a non-invasive method for monitoring cardiac function and other parameters of the cardiovascular system. Ann. N. Y. Acad. Sci. **170**, 724–732 (1970)
41. Lababidi, Z., Ehmke, D.A., Durnin, R.E., Leaverston, P.E., Lauer, R.M.: The first derivative thoracic impedance cardiogram. Circulation **41**(4), 651–658 (1970)
42. Lehr, J.: A vector derivation useful in impedance plethysmographic field calculation. IEEE Trans. Biomed. Eng. **19**, 156–157 (1972)
43. Lozano, D.L., Norman, G., Knox, D., Wood, B.L., Miller, B.D., Emery, C.F., Berntson, G.G.: Where to B in dZ/dt. Psychophysiology. **44**(1), 113–119 (2007)
44. Malmivuo, J., Plonsey, R.: Bioelectromagnetism. Principles and Applications of Bioelectric and Biomagnetic Fields. Oxford University Press, Oxford (1995)
45. Meijer, J.H., Boesveldt, S., Elbertse, E., Berendse, H.W.: Method to measure autonomic control of cardiac function using time interval parameters from impedance cardiography. Physiol. Meas. **29**(6), S383–S391 (2008)
46. Miles, D.S., Sawka, M.N., Hanpeter, D.E., Foster-Jr, J.E., Doerr, B.M., Frey, M.A.B.: Central hemodynamics during progressive upper- and lower-body exercise and recovery. J. Appl. Physiol. **57**(2), 366–370 (1984)
47. Milsom, I., Sivertsson, R., Biber, B., Olsson, T.: Measurement of stroke volume with impedance cardiography. Clin. Physiol. **2**, 409–417 (1982)
48. Miyamoto, Y., Takahashi, M., Tamura, T., Nakamura, T., Hiura, T., Mikami, T.: Continous determination of cardiac output during exercise by the use of impedance plethysmography. Med. Biol. Eng. Comput. **19**, 638–644 (1981)
49. Miyamoto, Y., Tamura, T., Mikami, T.: Automatic determination of cardiac output using an impedance plethysmography. Biotelem. Patient Monit. **8**, 189–203 (1981)
50. Miyamoto, Y., Hiuguchi, J., Abe, Y., Hiura, T., Nakazono, Y., Mikami, T.: Dynamics of cardiac output and systolic time intervals in supine and upright exercise. J. Appl. Physiol. **55**(6), 1674–1681 (1983)
51. Miyamoto, Y., Hiura, T., Tamura, T., Nakamura, T., Hiuguchi, J., Mikami, T.: Dynamics of cardiac, respiratory, and metabolic function in men in response to step work load. J. Appl. Physiol. **52**(5), 1198–1208 (1982)
52. Moene, R.J., Mook, G.A., Kruizinga, K., Bergstra, A., Bossina, K.K.: Value of systolic time intervals in assessing severity of congenital aortic stenosis in children. Br Heart J **37**, 1113–1122 (1975)
53. Mohapatra, S.N.: Noninvasive Cardiovascular Monitoring of Electrical Impedance Technique. Pitman, London (1981)
54. Mohapatra, S.N.: Impedance cardiography. In: Webster, J.G. (ed.) Encyclopedia of Medical Devices and Instruments, pp. 1622–1632. Wiley, New York (1988)
55. Moon, J.K., Coggan, A.R., Hopper, M.K., Baker, L.E., Coyle, E.F.: Stroke volume measurement during supine and upright cycle exercise by impedance cardiography. Ann. Biomed. Eng. **22**, 514–523 (1994)
56. Mortarelli, J.R.: A generalisation of the Geselowitz relationship useful in impedance plethysmographic field calculation. IEEE Trans. Biomed. Eng. **27**, 665–667 (1980)
57. Murgo, J.P., Westerhof, N., Giolma, J.P., Altobelli, S.A.: Aortic input impedance in normal man: relationship to pressure wave forms. Circulation **62**(1), 105–116 (1980)
58. Muzzi, M., Jeutter, D.C., Smith, J.J.: Computer-automated impedance-derived cardiac indexes. IEEE Trans. Biomed. Eng. **33**(1), 42–47 (1986)

59. Nakagawara, M., Yamakoshi, K.: A portable instrument for non-invasive monitoring of beat-by-beat cardiovascular haemodynamic parameters based on the volume-compensation and electrical-admittance method. Med. Biol. Eng. Comput. **38**(1), 17–25 (2000)
60. Nakonezny, P.A., Kowalewski, R.B., Ernst, J.M., Hawkley, L.C., Lozano, D.L., Litvack, D.A., Berntson, G.G., Sollers 3rd, J.J., Kizakevich, P., Cacioppo, J.T., Lovallo, W.R.: New ambulatory impedance cardiograph validated against the Minnesota Impedance Cardiograph. Psychophysiology **38**(3), 465–473 (2001)
61. Nowakowski, A., Wtorek, J., Stelter, J.: Technical University of Gdansk Electroimpedance mammograph. IX International Conference on Bio-Impedance, Heidelberg, pp. 434–437 (1995)
62. Nyboer, J., Bango, S., Barnett, A., et al.: Radiocardiograms. J. Clin. Invest. **19**, 773–778 (1940)
63. Nyboer, J.: Plethysmography. Impedance. In: Glasser, O. (ed.) Medical Physics, vol. 2, pp. 736–743, Year Book Publishers, Chicago (1950)
64. Pałko, T., Galwas, B.: Electrical properties of biological tissues, their measurements and biomedical applications. Automedica **17**, 343–365 (1999)
65. Patterson, R.P.: Sources of the thoracic cardiogenic electrical impedance signal as determined by a model. Med. Biol. Eng. Comput. **23**, 411–417 (1985)
66. Penney, B.C.: Theory and cardiac applications of electrical impedance measurements. CRC Crit. Rev. Bioeng. **13**, 227–281 (1986)
67. Pethig, R.: Dielectric properties of body tissues. Clin. Phys. Physiol. Meas. **8**(suppl. A), 5–12 (1987)
68. Pickett, B.R., Buell, J.C.: Usefulness of the impedance cardiogram to reflect left ventricular diastolic function. Am. J. Cardiol. **71**(12), 1099–1103 (1993)
69. Porter, J.M., Swain, I.D.: Measurement of cardiac output by electrical impedance plethysmography. J. Biomed. Eng. **9**(3), 222–231 (1987)
70. Quail, A.W., Traugott, F.M.: Effects of changing haematocrit, ventricular rate and myocardial inotropy on the accuracy of impedance cardiography. Clin. Exp. Pharmacol. Physiol. **8**(4), 335–343, (1981)
71. Raaijmakers, E., Faes, J.C., Goovaerts, H.G., de Vries, P.M.J.M., Heethaar, R.M.: The inacuracy of Kubicek's one-cylinder model in thoracic impedance cardiography. IEEE Trans. Biomed. Eng. **44**(1), 70–76 (1997)
72. Riese, H., Groot, P.F., van den Berg, M., Kupper, N.H., Magnee, E.H., Rohaan, E.J., Vrijkotte, T.G., Willemsen, G., de Geus, E.J.: Large-scale ensemble averaging of ambulatory impedance cardiograms. Behav. Res. Methods Instrum. Comput. **35**(3), 467–477 (2003)
73. Rosenberg, P., Yancy, C.W.: Noninvasive assessment of hemodynamics: an emphasis on bioimpedance cardiography. Curr. Opin. Cardiol. **15**(3), 151–155 (2000)
74. Sakamoto, K., Muto, K., Kanai, H., Iiuzuka, M.: Problems of impedance cardiography. Med. Eng. Comput. **17**, 697–709 (1979)
75. Seigel, J.H., Fabian, M., Lankau, C., Levine, M., Cole, A., Nahmad, M.: Clinical and experimental use of thoracic impedance plethysmography in quantifying myocardial contractility. Surgery **67**, 907–917 (1970)
76. Shankar, T.M.R., Webster, J.G.: Automatically balancing electrical impedance plethysmography. J. Clin. Eng. **9**, 129–134 (1984)
77. Shankar, T.M.R., Webster, J.G., Shao, S.-Y.: The contribution of vessel volume change to the electrical impedance pulse. IEEE Trans. Biomed. Eng **1**, 42–47 (1986)
78. Sherwood, A., McFetridge, J., Hutcheson, J.S.: Ambulatory impedance cardiography: a feasibility study. J. Appl. Physiol. **85**(6), 2365–2369 (1998)
79. Siebert, J.: Impedance cardiography—atrio-ventricular heart blocs (in Polish). Elektrofizjologia i Elektrostymulacja Serca **2**, 28–34 (1995)
80. Siebert, J., Wtorek, J.: Impedance cardiography. The early diastolic phase of heart chambers feeling—O dz/dt. (in Polish). Ann Acad Gedan **23**, 79–89 (1993)
81. Smith, J.J., Bush, J.E., Wiedmeier, V.T., Tristani, F.E.: Application of impedance cardiography to study of postural stress. J. Appl. Physiol. **29**(1), 133–137 (1970)

82. Smith, J.J., Muzi, M., Barney, J.A., Ceschinn, J., Hayes, J., Ebert, T.J.: Impedance-derived cardiac indices in supine and upright exercise. Ann. Biomed. Eng. **17**(5), 507–515 (1989)
83. Smorawiński, J., Nazar, K., Kaciuba-Uscilko, H., Kamińska, E., Cybulski, G., Kodrzycka, A., Bicz, B., Greenleaf, J.E.: Effects of 3-day bed rest on physiological responses to graded exercise in athletes and sedentary men. J. Appl. Physiol. **91**, 249–257 (2001)
84. Sramek, B.B.: Cardiac output by electrical impedance. Med. Electron. **13**(2), 93–97 (1982)
85. Sramek, B.B.: Hemodynamic and pump-performance monitoring by electrical bioimpedance: new concepts. Probl. Resp Care **2**(2), 274–290 (1989)
86. Sramek, B.B.: Thoracic electrical bioimpedance: Basic principles and physiologic relationship. Noninvas Cardiol **3**(2), 83–88 (1994)
87. Swanson, D.K., Webster, J.G.: Errors in four-electrode impedance plethysmography. Med. Biol. Eng. Comput. **21**(6), 674–680 (1983)
88. Takada, K., Fujinami, T., Senda, K., Nakayama, K., Nakano, S.: Clinical study of "A waves" (atrial waves) in impedance cardiograms. Am. Heart J. **94**(6), 710–717 (1977)
89. Tanaka, K., Kanai, H., Nakayama, K., Ono, N.: The impedance of blood: the effects of red cell orientation and its application. Jpn. J. Med. Eng. **8**, 436–443 (1970)
90. Thomsen, A.: Impedance cardiography—Is the output from right or from left ventricle measured ? Intensive Care Med. **6**, 206 (1979)
91. Yamamoto, Y., Mokushi, K., Tamura, S., Mutoh, Y., Miyashita, M., Hamamoto, H.: Design and implementation of a digital filter for beat-by-beat impedance cardiography. IEEE Trans. Biomed. Eng. **12**, 1086–1090 (1988)
92. Webster, J.G. (ed.): Medical Instrumentation. Application and Design. Wiley, New York (2010)
93. Willemsen, G.H., De Geus, E.J., Klaver, C.H., Van Doornen, L.J., Carroll, D.: Ambulatory monitoring of the impedance cardiogram. Psychophysiology **33**(2), 184–193 (1996)
94. Winter, U.J., Klocke, R.K., Kubicek, W.G., Niederlag, W. (eds): Thoracic Impedance Measurements in Clinical Cardiology. Thieme Medical Publishers, New York (1994)
95. Wtorek, J., Poliński, A., Stelter, J., Nowakowski, A.: Cell for Measurements of Biological Tissue Complex Conductivity, Technology and Health Care, vol. 6, pp. 177–193. IOS Press (1998)
96. Wtorek, J., Poliński, A.: The contribution of blood-flow-induced conductivity changes to measured impedance. IEEE Trans. Biomed. Eng. **52**(1), 41–49 (2005)
97. Zhang, Y., Qu, M., Webster, J.G., Tompkins, W.J.: Impedance cardiography for ambulatory subjects. In: Proceedings of the Seventh Annual Conference of the IEEE/Engineering in Medicine and Biology Society. Frontiers of Engineering and Computing in Health Care, vol. 2, pp. 764–769. IEEE, New York (1985)
98. Ziemba, A.W., Chwalbińska-Moneta, J., Kaciuba-Uscilko, H., Kruk, B., Krzeminski, K., Cybulski, G., Nazar, K.: Early effects of short-term aerobic training. Physiological responses to graded exercise. J. Sports Med. Phys. Fitness **43**(1), 57–63 (2003)
99. Web page of Medicare and MediAid: http://new.cms.hhs.gov/manuals/downloads/Pub06_PART_50.pdf

Chapter 3
Ambulatory Impedance Cardiography

In this chapter the idea of the impedance cardiography will be presented followed by a description of a central hemodynamics ambulatory monitoring system, available for both research and clinical applications. Basing on the REOMONITOR system, developed in my home institution, the essential parts of the most of the ambulatory systems will be presented. Then will be described other systems used in the cardiology and physiological researches. All of the presented in this chapter systems were verified using the clinically accepted reference methods—non-invasive (ultrasound), invasive or at least, compared against the stationary impedance device Minnesota 304B.

3.1 The Idea of Ambulatory Impedance Cardiography

The idea of impedance cardiography ambulatory monitoring with signal recording on memory chips was introduced in 1985 and 1987 by Webster's group [17, 21], and developed using PCMCIA cards [4, 5, 6, 8, 9, 10, 11]. In 1996, the ambulatory monitoring of an impedance cardiogram (ICG) device was described and the results were collected for 26 subjects during 24-h monitoring, and for 25 subjects in various conditions [20]. The device used, however, enabled only the storage of the results of calculations performed on ensemble-averaged signals so it did not give access to every single heartbeat. Sherwood et al. [19] have also, published a paper revealing the existence of his ambulatory impedance monitor (AIM) device. The primary objective of their study was "to assess how a newly developed AIM would compare with established impedance cardiographic instrumentation". However, "the study's objective was not to revisit the validation of impedance cardiography as a technique per se but rather to evaluate whether the AIM system with its unique hardware and electrode configuration, would yield cardiovascular function indexes similar to those that are obtained with standardised impedance cardiographic methodology." In 2000, a portable poly-physiograph for non-inva-

G. Cybulski, *Ambulatory Impedance Cardiography*,
Lecture Notes in Electrical Engineering, 76, DOI: 10.1007/978-3-642-11987-3_3,
© Springer-Verlag Berlin Heidelberg 2011

sive monitoring of beat-by-beat cardiovascular hemodynamic parameters based on the volume-compensation and electrical-admittance method was described [13]. Their portable unit is able to control measurement procedures, performs blood pressure and cardiac output (CO) measurement, processes signals and stores almost 32,000 beats of time-series data in a fully automated manner. The device was used for evaluation of a subject's cardiovascular hemodynamic responses to daily physical activities as well as to various psycho-physiological stresses. Nakonezny et al. [14] compared the results obtained using their device with those recorded using stationary equipment during rest and some behavioural challenges in the laboratory. They found ambulatory ICG to be a reliable method for measurement of stroke volume (SV), CO, heart rate (HR) and systolic time intervals during a variety of psycho-physiological tests. Another group [15] presented the low-power consumption system allowing the wireless Bluetooth® transmission of the signals. Recently, some commercial devices have been introduced: the MindWare 1000A (MW1000A) (http://www.mindwaretech.com) ambulatory impedance cardiograph, for example, is composed of two parts—a small battery-powered instrument and a palmtop-based data acquisition system. In 2006 Parry and McFetridge-Durdle [16] performed a systematic review of three ambulatory impedance cardiography devices. They used AIM, VU-AMS, and AZCG systems to examine cardiovascular indices. Although they spotted some inconsistencies in determining pre-ejection period (PEP) and SV the validity, reliability and sensitivity of the analyzed systems seemed to be acceptable. The conclusion of that paper was that "availability of non-invasive ambulatory measures of cardiac function has the potential to improve care for variety of patient populations, including those with hypertension, heart failure, pain, anxiety, and depressive symptoms". Following the studies using stationary impedance cardiography in hypertension therapy monitoring this type of the clinical application seems to be very promising when ambulatory technique will be applied [12].

Some of the constructions described below passed the necessary certification processes and could be available on the market as commercial systems for clinical applications, whereas other reminded only in the research domain.

3.2 ReoMonitor: The Research System

In 1993 the author with his team started the development of the system for ambulatory monitoring of impedance signals with the data storage on PCMCIA cards. Initially the highest offered capacity of that type of the memory was at the level of 20 MB, which seems now very small. At those times it allowed for more than 12 h recording of two-channel signals at the sampling frequency of 200 Hz. The system uses the tetrapolar configuration of the spot electrodes. Figure 3.1 presents the configuration of the electrodes on patient during the in-lab recordings. Outside the laboratory the "forehead electrode" is placed on the neck above the receiving electrode.

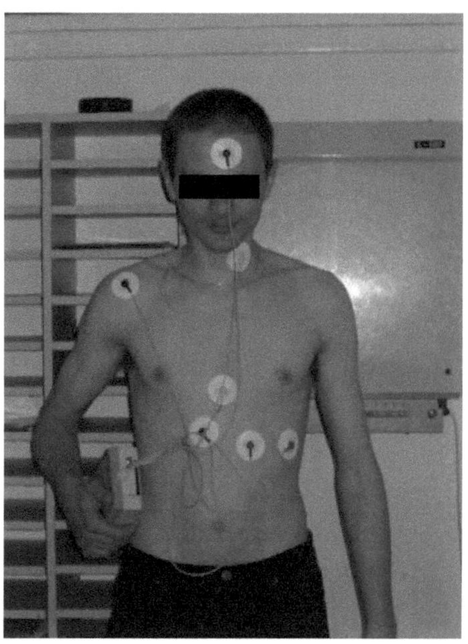

Fig. 3.1 The configuration of the electrodes on patient during the in-lab recordings using ReoMonitor

3.2.1 The Ambulatory Recorder

ReoMonitor, the central hemodynamic ambulatory recording device is composed of analogue (signal detecting) and digital (data recording) units. Both units were constructed using high quality low noise industrial grade integrated circuits. Their technical data are presented in Appendix.

Figure 3.2 presents ReoMonitor, the research long-time ambulatory monitoring system developed at Medical Research Centre (Warsaw, Poland) in collaboration with the Institute of Precise and Biomedical Engineering at Warsaw University of Technology (Poland).

3.2.2 The Analogue Unit

A new miniaturised, tetrapolar, current impedance cardiograph with one built-in electrocardiogram (ECG) channel was designed and constructed. The block scheme is presented in Fig. 3.3. A set of combined band and point type electrodes positioned in slightly modified electrode configuration was used for the impedance cardiography.

The application generators produce a stabilised 95 kHz sinusoidal voltage signal that is converted to a current signal (high output resistance) and applied to the subject's chest via a pair of application electrodes (A). The voltage signal detected on receiving electrodes (R) is amplified and demodulated in the peak detector.

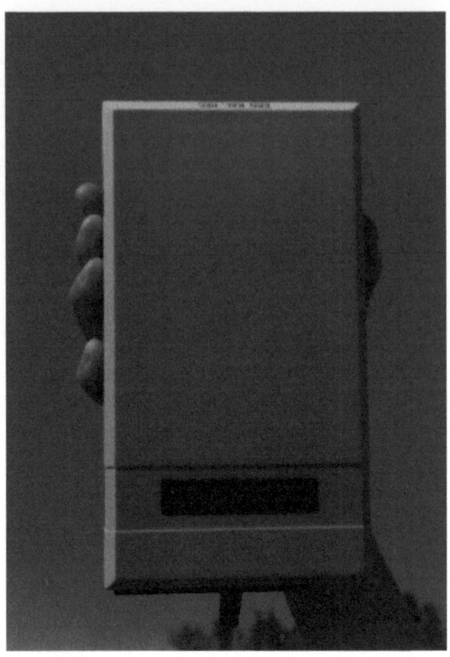

Fig. 3.2 Ambulatory monitoring recorder (ReoMonitor) for long-time data acquisition in research applications

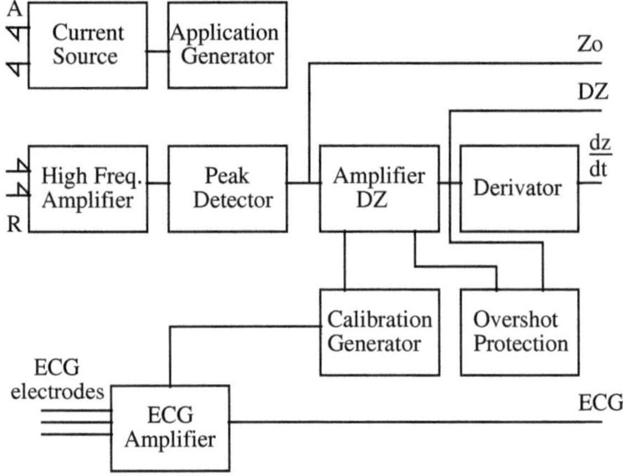

Fig. 3.3 Block scheme of the miniaturised impedance tetrapolar cardiograph with built-in one channel of ECG. Three electrodes are used to collect the one channel ECG and four electrodes are used to generate impedance cardiography signals

Application of an LC coupled generator with a low noise active element powered with stabilised voltage ensures the high long-term stability of both amplitude and frequency. This feature is fundamental for detection of high quality ICGs. The electric signal from generator is transmitted to the input of the operation amplifier, which works as a converter of alternating voltage to alternating current of constant amplitude. The output impedance of this current generator is 100 kΩ. The load is the primary winding of the ferromagnetic core transformer. The ends of the secondary winding are connected to the application electrodes. The receiving electrodes are connected to the symmetrical input of the instrumentation amplifier (input resistance >200 MΩ). The amplified signal passes the peak-detection module with the aim of obtaining the ICG signal. The signal is then decomposed to a constant component that reflects the signal of basic impedance (Z_0) and a fluctuating part of the signal (ΔZ) that reflects changes in the volume of blood in the chest. This signal passes through two-stage analogue derivation with the properly selected characteristic [linear amplification within the bandwidth up to 15 Hz (3 dB), giving the gain of 1 V/Ω/s for dz/dt signal]. A specialised overshoot protection circuit has been developed with the aim of shortening any long-term voltage saturation when artefacts occur instead of a clear signal. The gains of amplifiers are set to obtain the sensitivity of output signals for ΔZ and Z_0 at, respectively, about 1 V/100 mΩ and 1 V/Ω. Calibration signals are provided by specialised module ($Z_0 = 20$ Ω, dz/d$t = 1$ V/Ω/s, $\Delta Z = 50$ mΩ, and $a = 1$ mV for ECG channel). Both analogue and digital parts of the device are powered by 6×1.5 V alkaline AA (R6) type batteries. The voltage is transformed to the demanded levels using DC/DC converters and stabilised.

3.2.3 The Digital Unit

The recording part of the device (block scheme presented on Fig. 3.4) is based on an 80C552 family microcontroller with built-in four 8-bit A/D converters. The signals, ECG, the first derivative of the impedance signal (dz/dt), changes in the impedance (ΔZ), and the value of the basic thoracic impedance (Z_0), change within the range of 0–5 V, which was chosen for this application. They are sampled at the rate of 200 Hz. In the prototype, the specialised assembler procedure is stored in EPROM 27C128. For temporary data storage and necessary signal analysis a 256 kbit RAM

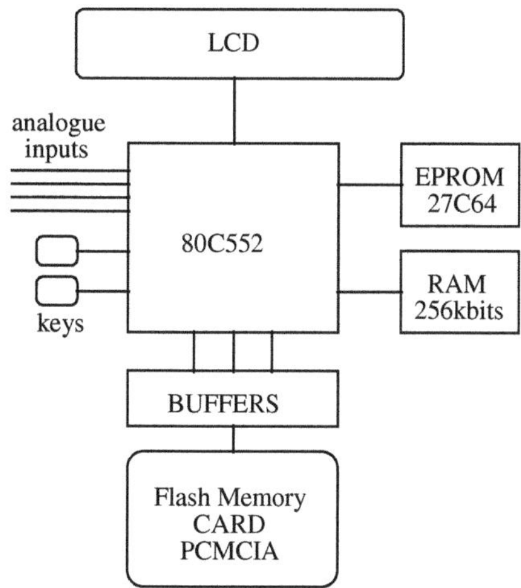

Fig. 3.4 Block scheme of the digital part of hardware. The analogue inputs are feed by ECG, dz/dt, ΔZ, and Z_0 from the detecting part of the recorder

is used. Data are stored on a 20 MB Flash Memory Card prepared according to the Standard of PCMCIA v.2.01 type II and working in the Memory Mapped mode. New models of PCMCIA cards (or Compact Flash Cards with PCMCIA adapters) allow for even greater data storage—up to 512 MB. Communication with the system is performed via specialised keys and a small, built-in alphanumerical LCD. Special procedures have been prepared to save the power consumption.

3.2.4 The User Interface

There are two parts of the graphic user interface which works in a Windows environment: ReoMake and ReoMon. Prior to initialisation of the ReoMon display interface module, ReoMake must be run to extract data from the card, introduce patient and examination parameters and convert raw data from the recorder into the format used by a display module. ReoMake enables selection of source and destination files, sampling frequency and number of channels and running of the calculation of hemodynamics parameters procedure. The ReoMake opening screen is presented in Fig. 3.5.

ReoMon allows for data presentation in a bioscope and a tape strip (TS) for 1, 2, 3 or 4 independent channels. Additionally, full disclosure (FD) data (1 h of recording of a channel per A4 page, 1 min of the recording per line) and selected

Fig. 3.5 Opening screen of the graphic user interface of ReoMake

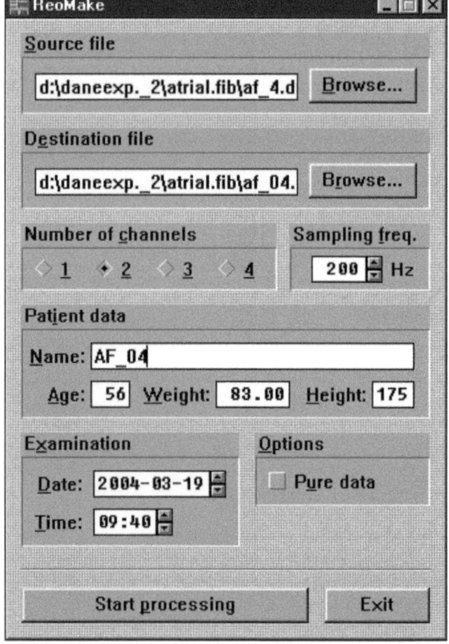

strips may be hard copied to a laser or ink-jet printer. To allow more detailed presentation the signals may be shown at the following standard speeds: 10, 25, 50, 100 mm/s and FD (1 min per line) and multiplied by the following factors: 0.5, 1.0, 1.5, and 2.0. Also pre-scaling is possible according to the geometric progression: 1/8, 1/4, 1/2, 1, 2, 4, and 8. Changes between these functions may be made by both pull-down menu and specialised buttons. Search tools also allow for finding markers or particular times of registration. In Options, the colour of traces and grid lines may be modified, as may be the level of calibration, the position of the neutral line and the patient's data. ReoMon has the following tools allowing for efficient browsing and analysis of the ICG traces:

1. time selection using scroll bar, insertion of a number into dialog box or jumping to the nearest marker,
2. scales showing the cursor's position on the time and amplitude axes in the respective units,
3. measurement between two different points on a selected channel using a calliper function.

For FD and TS presentation modes it is possible to select the scale, page format, raster and resolution level. Sample printouts of this program are used as figures in this thesis. The ReoMon opening screen is presented on Fig. 3.6.

3.2.5 Software for Hemodynamics Parameters Calculations

A specialised program written in C language analyses the collected data and enables automatic determination of cardiac parameters. This program consists of the following procedures: detection of the QRS complex, detection of characteristic points on the dz/dt ICG curve, reduction of the breathing artefacts, creation of the results matrix, artefact rejection subroutine. The first two procedures find the characteristic points on ECG and ICG curves. These points are Q on QRS complex, the point of crossing the baseline by the ICG signal, the maximum of the dz/dt_{max} signal, and the closing of the aortic valve point. A specially designed procedure for reducing breathing artefacts enables and performs a low pass filtering using both the ECG and ICG signals [3]. On the basis of these data the following parameters are calculated for each cycle: time of the cycle occurrence (T), HR, SV, CO, distance between two QRS complexes (RR), ejection time (ET), PEP, maximal amplitude of ICG signal (dz/dt_{max} = AMP) and basic impedance of the chest (Z_0). Each cycle is also classified as a normal (N) or artefact (A). Additionally the mean values of each parameter and their standard deviation are calculated over the period selected by the operator. As an example the printout obtained from a 25-year-old healthy subject is presented in Table 3.1. Zero in the last column (N/A) denotes detection of an artefact; one denotes that the cycle was classified as normal [3].

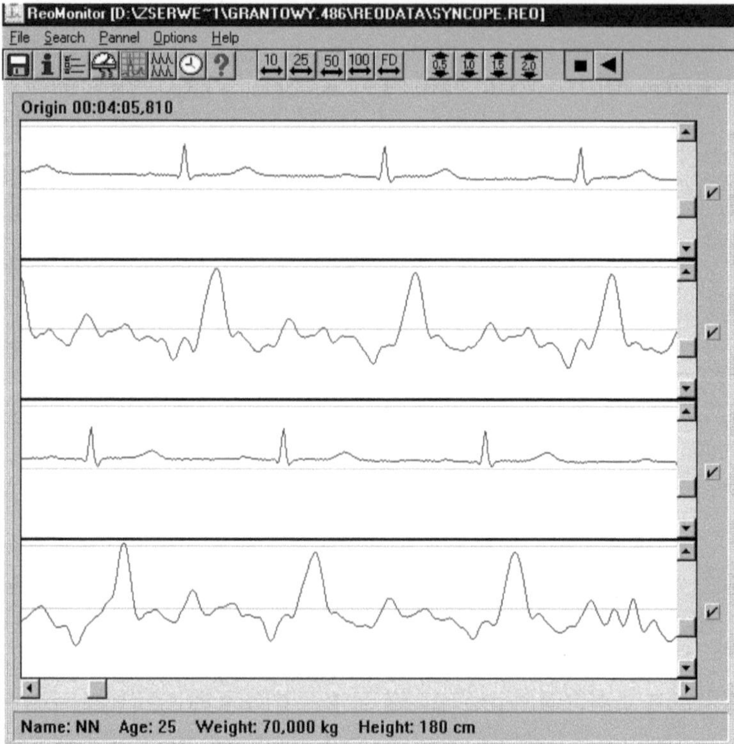

Fig. 3.6 Opening screen of the graphic user interface of REMONITOR

Table 3.1 Beat-to-beat variations in hemodynamic parameters in a healthy subject (25 years)

Report

Time (s)	HR (1/min)	SV (ml)	CO (l/min)	RR (ms)	ET (ms)	PEP (ms)	$(dz/dt)_{max}$ (Ω/s)	Z_0 (Ω)	*N/A*
01:51	61	56	3.4	975	215	120	2.59	26.0	1
01:52	64	61	3.9	930	260	110	2.34	26.0	1
01:53	61	53	3.2	970	230	120	2.29	26.0	1
01:54	59	55	3.2	1,010	235	130	2.31	26.0	1
01:55	61	84	5.1	980	305	115	2.72	26.0	1
01:56	60	61	3.7	1,000	265	120	2.27	26.0	1
01:57	56	65	3.6	1,055	295	130	2.18	26.0	1
01:58	59	39	2.3	1,015	160	135	2.40	26.0	0
01:59	61	86	5.2	970	330	105	2.59	26.0	1
Mean									
02:00	60	66	4.0	988	272	118	2.39	26.0	$N = 9$
SD	4	12	0.8	47	44	13	0.22	0.00	$A = 1$

A recognised artefact is shown as 0 in the last column (*N/A*)

3.3 VU-AMS: The Vrije Universiteit Ambulatory Monitoring System

The Ambulatory Monitoring System (VU-AMS) was developed at the Vrije Universiteit in Amsterdam. For the first time it was applied to collect data from for 26 subjects during 24-h monitoring, and for 25 subjects in various conditions [20]. This battery-powered Holter-type device, is capable of recording a three lead ECG and a two lead skin conductance level (SCL) (version 3.6) or a four lead ICG (version 4.6). The image of the Ambulatory Monitoring System (VU-AMS) could be found at the following web page: http://www.psy.vu.nl/vu-ams/information/index.html).

The VU-AMS is used in several psycho-physiological laboratories in stationary applications, but its full potential is realised in ambulatory research. It is especially important in psychophysiological researches that should be not affected by an artificial laboratory situation. The VU-AMS is small (dimensions $32 \times 65 \times 120$ mm), weighing only 225 g. Despite of the miniaturisation, reliability and validity of the ECG and ICG signals has remained very high, which resulted in the numerous research performed and papers published in well-recognised journals. The system is used worldwide by 38 research groups (according to the web page data, see the link: http://www.psy.vu.nl/vu-ams/research/index.html) to study stress and emotion in the natural environment of the subject. Over 35 publications now feature the VU-AMS that would have been difficult if not impossible without this ambulatory technology. The list of papers containing the research results obtained using VU-AMS is available under the following link: http://www.psy.vu.nl/vu-ams/research/publications.html.

The device is intensively used in the ongoing cardiovascular stress and genetics research of the department of Biological Psychology and the Netherlands Twin Registry. The developers of the system claims that "(…) ongoing scientific demand from within has maintained the VU-AMS (…)" creates the "(…) continuous state of innovation since its first prototypes in the early 1990s (…)". The current version is marked "5fs". The system is offered in two versions: the ICG version (VU-AMS46) and the SCL version (VU-AMS36), which does not possess the ICG unit (and is out of the scope of that monograph). The ICG version measures ECG, ICG and motility (the ability to move spontaneously and actively, consuming energy in the process) and the SCL version measures ECG, SCL and motility. Optionally the ICG version could be fortified with Phonocardiogram channel or Individual Noise Exposure measurement. The VU-AMS can be used to assess autonomic reactivity during stress and conditioned emotional responses. Measured variables are heart period (HP), HP variability (SDNN, RMSSD indices), PEP, respiration rate (RR), respiratory sinus arrhythmia (RSA), SV and CO, SCL, vertical acceleration (gross body movement).

The VU-AMS uses six spot electrodes to receive ECG and ICG. Two electrodes are place on the back and are used as a high-frequency application points. On the chest are placed two electrodes receiving the voltage difference over the thorax. Two other electrodes are placed on the both sides of the lower part of the chest. The topology of the electrodes placement could be found in the VU-AMS manual (http://www.psy.vu.nl/vu-ams/manuals/index.html).

Below is a quotation from that manual describing the method of the impedance signal generation an analysis. "The thorax impedance measurement uses a 4-spot electrode technique. Two electrodes supply a constant current through the subject of 50 kHz, 350 µA. The remaining two electrodes pick up the impedance signal. This signal is amplified and led to a precision rectifier. The rectified signal is filtered at 72 Hz to give thorax impedance Z_Q. Filtering Z_Q at 0.1 Hz supplies $A\dot{Z}$, which in return is filtered at 30.0 Hz to determine dz/dt. Z_0, ΔZ and dz/dt are led to the AD-converter of the microprocessor. The ΔZ is sampled default at 100 ms yielding the respiration signal. This sample rate can be set between 100 and 1,000 ms. AMS stores the respiration signal for later off-line processing. This off-line processing can be used to obtain inspiration time, expiration time and total cycle time off all recorded breaths. Signal dz/dt is sampled at 250 Hz to yield the ICG. dz/dt values are sampled only during a short period (512 ms) around each R-wave. These R-wave locked dz/dt data blocks are ensemble averaged over a default period of 60 s. The user may change this from 30 to 120 s. The entire ensemble average is written to AMS memory, including the average Z_0 value measured at the beginning of the block. Off-line processing of the ensemble averages of the dz/dt signal allows the computation of systolic time intervals like the left ventricular ejection ctime (LVET), the PEP and the Heather index" (http://www.psy.vu.nl/vu-ams/manuals/index.html). VU-AMS does not record the full 3-channel ECG. Instead a series of R–R intervals is stored [20]. The R-wave is used as a trigger allowing the synchronisation necessary to perform the ICG signal averaging. The ensemble averaging is performed over a certain period (e.g. 60 s) by summing the digitised samples gated by R-wave peak and dividing by the numbers of beats in the analysed period. It is used to reduce the influence of natural beat-to-beat variability and the disturbing effect of the respiration on impedance signal.

Another method used in analysis of the ambulatory ICG is the large scale ensemble averaging. This method is a way of longer averaging than over the 60 s period. It was introduced by the group who developed the VU-AMS device [18]. This method was applied as an off-line analysis for a period of the similar type of the patient activity but lasting not longer than 1 h. If the activity type lasted longer than 1 h (sleeping) it was divided into several periods shorter than 1 h. The periods were selected using the entries from patient diaries describing the activity, physical load, posture, location (home, work, etc.), and social situation. The full description of the categories is given by Riese et al. [18]. The morphology analysis of the large-scale ensemble averaging was performed in a similar way to a 60 s averaging. The technical specification data are presented in Appendix.

3.4 MW1000A: The MindWare System

The MindWare MW1000A Ambulatory Impedance Cardiograph (Mindware Technologies LTD, Gahanna, OH 43230, USA, http://mindwaretech.com) is a battery powered, portable, small impedance device combined with a PDA based data acquisition system. The impedance signal detector is design as a tetra-polar system using two application and two receiving electrodes. The application part of MW1000A injects a high precision frequency alternating current (l00 kHz) of constant amplitude at 400 mA. This allows measurement of the impedance changes across the thorax. The detection part contains an ECG receiving and amplification channel. The four continuous output signals—impedance (Z_0), its derivative (dz/dt), ECG, and a galvanic skin conductance (GSC)—are digitized by the high-resolution (16 bits) A/D card. Data can be streamed to local storage on the PDA (512 MB SD) or transmitted Wi-Fi to a host PC using a wireless router and MindWare's Wi-Fi acquisition application (ACQ). This data can then be analyzed real-time (using so called ICG RT) or offline with MindWare's suite of analysis applications. Real time measurements are stored in a tab delimited text file for offline processing. File format is compatible with MindWare desktop analysis applications IMP, HRV, and EDA. The system has very low power consumption and is powered by internal PDA battery. The device of a size $45 \times 95 \times 160$ mm weights 400 g. The system enables to compute and display real time cardiac and systolic measures such as: LVET, PEP, SV, CO, HR, dz/dt, Z_0, mean inter beat interval (IBI) and RSA for hear rate variability, and skin conductance.

The system is supplied complete with the PDA, rugged case with Impedance Electronics, Acquisition Card, SD Card (for memory storage), patient leads and harness, charger, and acquisition software. The wireless router essential to use the included Wi-Fi ACQ acquisition software should be bought separately. The technical data and features are presented in the appendices chapter. Figure 3.7 presents the MindWare MW1000A Ambulatory Impedance Cardiograph (Mindware Technologies LTD, Gahanna, OH 43230, USA). Some technical data of that device are presented in Appendices part.

3.5 PhysioFlow Enduro System

The PhysioFlow Enduro is a Holter-type version of a system offered also by the producer in a stationary version. It is very light (200 g with AA batteries) and offers a long time recording in a field conditions. The crucial difference between this system and other available on the market is associated with the way of calculating the SV from an impedance signals. It is the only system that does not use geometrical parameters of the thorax (or a distance between the electrodes) to calculate every cycle SV values. Due to some legal and commercial reasons the method of calculation was never revealed in the scientific paper. According to the

Fig. 3.7 The MindWare
MW1000A Ambulatory
Impedance Cardiograph
[Mindware Technologies
LTD, Gahanna, OH 43230,
USA (reproduced from the
web page
http://mindwaretech.com
with permission)]

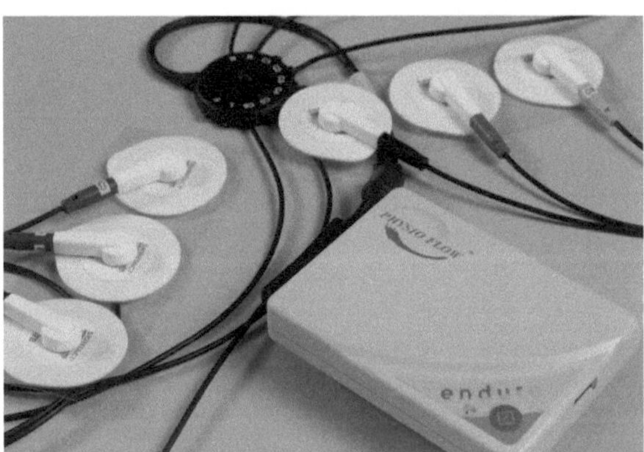

Fig. 3.8 The PhysioFlow
Enduro recorder (reproduced
from the materials received
from the PhysioFlow Enduro
producer with permission)

information from a producer and the general description it could be concluded that calculations are based only on shape and parameters of the impedance signals. The foundations of SV calculations were presented in Appendix 1 of the paper [1] and described earlier in the chapter where SV formulas were presented. Figure 2.6 presents the topology of the electrodes position on the thorax. The closer view of the PhysioFlow Enduro showing the size of the recording device is presented on the Fig. 3.8.

To generate the transthoracic impedance signal the system uses the tetrapolar method. The Physio Flow emits a high-frequency (75 kHz) and low-amperage (1.8 mA) alternating electrical current via electrodes. Two sets of two electrodes (Ag/AgCl, Hewlett Packard 40493 E), one "transmitting" electrode, one "sensing" electrode, are applied above the supraclavicular fossa at the left base of the neck and along the xiphoid, respectively. Positioning of the electrodes is not critical, since Z_0 evaluation is unnecessary. No specific skin preparation is needed, except shaving. Another set of two electrodes is used to monitor a single ECG lead (VI/V6 position). Verification of the correct signal quality is accomplished by visualization of the ECG and its first derivative (dECG/dt), the impedance waveform (ΔZ) and its first derivative (dz/dt). Autocalibration is achieved after having entering the patient's age, height, body mass, and systolic/diastolic blood pressure assessed using a standard mercury-column sphygmomanometer. To achieve this autocalibration, which provides the basic curves and data necessary to measure SV variations AC-cording to $A\dot{Z}$ and dz/dt variations, patients have to be immobile and relaxed. Values for SV obtained over a 12-beat period are averaged, the Physio Flow deleting unacceptable curves (Olympic filter level 2). For this experiment, CO determinations made with Physio Flow during 1 min were averaged.

The following parameters could be monitored using PhysioFlow Enduro version:

- Stroke volume CO/index contractility
- Ventricular ET ejection fraction (estimate)
- Early diastolic function ratio
- Systemic vascular resistance
- Left cardiac work index

The system, according to the producer, could be applied in the following fields:

- Cardiology
- Stress testing
- Physiology and sports medicine
- Military and aerospace medicine

Some technical data of that device are presented in Appendices part. With the aim to obtain data regarding the clinical applications of stationary PhysioFlow system (according to their information used by more than 240 centers worldwide) please visit their web page: http://www.physioflow.com/clinicalinformation.htm. PhysioFlow Enduro is waiting for FDA clearance (July 2010).

3.6 AIM-8-V3: Wearable Cardiac Performance Monitor

In 1998, Sherwood et al. [19] introduced their "AIM" device. It is also known as "Ambulatory Bioelectric Impedance Monitoring for Assessing Cardiac Performance". This note was prepared basing on their web page information and the scientific papers [12, 16, 19]. According to the designers, the battery-powered AIM-8-V3 was created to obtain cardiac performance information from human and animal subjects during ambulatory conditions. It contains an internal computer and a bioelectric impedance cardiograph. The monitor has the capability of acquiring and processing real-time impedance cardiography signals and computing the cardiac performance indices. These parameters include HR, SV, CO, systolic time intervals (PEP, LVET, and TEP), contractility, basal impedance (Z_0), $(dz/dt)_{max}$, patient activity, etc.

It can also record gross body activity during each ensemble-averaged measurement, which it then saves along with the impedance data. The AIM-8-V3 also stores the time and date that each measurement was obtained. The system can obtain and store up to 100 cardiac performance measurements and computed cardiac indices along with the ensemble-averaged waveforms for ECG, dz/dt, and Z_0. Also, the Data Scan information, which contains the data pertinent to each cardiac cycle comprising each ensemble-averaged measurement, is recorded and uploaded along with the data for future use during editing.

The AIM-8-V3 has very few connectors and controls. It has a 9-pin connector for the impedance electrode assembly, a phone plug for the RS-232 serial communications cable, a RCA connector for the remote blood pressure start, a manual start push button, and a red LED (which indicates when impedance data is being collected during an ensemble average by flashing when the R-Wave of the ECG is detected for each cardiac cycle) (http://www.microtronics-bit.com/BIT/Products.html).

The AIM-8 monitoring device consists of a small ($76 \times 100 \times 38$ mm) plastic enclosure that contains a small (credit card size) bioelectric impedance cardiograph (labeled the "WHIC8" for Wearable Hutcheson Impedance Cardiograph), a small (credit card size) microcomputer/data logger with sleep-wake capability, a body-activity sensor, a 9-V battery, and an electrode assembly consisting of five electrodes.

The AIM-8 is low power consumption device—powered from a standard (snap-connector type) 9-V alkaline battery. Its main advantage over other constructions was that AIM-8 was capable of both data acquisition as well as the analysis of the acquired data. Other attempts at developing a wearable CO monitoring device based upon impedance cardiography techniques has resulted in prototype systems that only acquire the data (at best), and required additional data processing and analysis by another computer system in order to determine he cardiac performance results. The AIM-8 provides this information as the data is collected. The device can acquire up to 200 Ensemble Averages including both Computed Indices as well as the Ensemble Averaged Waveforms. Since the device has a low-power sleep mode, consuming 200 µA from the battery, it can acquire the cardiac data over long time-intervals.

Potential applications of the AIM-8 device (quoted according to the producers):

- Assessment of the cardiac performance in ambulatory patient applications for clinical diagnosis purposes.
- In medical research studies involving the measurement of cardiac performance during both laboratory-based and ambulatory patient situations.
- In cardiovascular research studies involving the measurement of systolic time intervals over a number of cardiac cycles.
- When used in conjunction with a blood pressure monitor, the AIM-8 device is useful for the assessment of total peripheral resistance (TPR).

- For both clinical and laboratory applications involving real-time estimates of changes in a patient's CO.
- For long-term assessment of cardiac performance changes in ambulatory patients.
- For both research and clinical applications involving the study of cardiac performance relative to gross body activity.
- In blood pressure assessment applications by giving a more complete picture of the cardiovascular system's status relative to the blood pressure measures.

Some technical data of that device are presented in Appendices part.

According the information from the producer the production of this type is discontinued although in April 2010 a several copies were still available.

3.7 Ambulatory Impedance Cardiograph: AZCG

In 2001, Nakonezny et al., published the paper in which validated their ambulatory impedance cardiography called AZCG (World Wide Medical Instruments) against the Minnesota Impedance Cardiograph. Basing on that paper we can learn that the AZCG is a wearable recorder designed for non-invasive acquisition of physiological data during daily activity. It has a size of $45 \times 95 \times 160$ mm and weight of 400 g with batteries. The analogue subsystem comprises a three-lead ECG (bandwidth 0.05–100 Hz) and a four-lead electrical impedance system (tetrapolar), which provides an alternating current of constant amplitude (2 mA RMS at 50 kHz) to the outer electrodes and records thoracic electrical impedance via the inner electrodes. The acquired analogue impedance signal is filtered, amplified and differentiated to produce signals for Z_0 (DC 100 Hz), DZ (DC 40 Hz), and dz/dt (DC 40 Hz). The ECG and ICG each employ a digitally controlled, sampled-signal rebalance method for waveform stability. The digital subsystem provides A–D conversion of the above signals using a Motorola MC68332-based microcomputer with a 12-bit A–D converter and 256 kB RAM. Digitized signals are stored on a 20 MB Flash Card (PCMCIA), allowing 30 min of data acquisition at the 500 Hz A–D sampling rate. Programming during set up, signal monitoring, and uploading of data are accomplished using standard communication software through digital input/output connectors and a serial interface to a microcomputer system.

Ambulatory data acquisition is controlled using an onboard programmable protocol manager. User selectable protocol parameters include A–D sampling rate (100–1,000 Hz), analogue channel selection, and timed or triggered initiation of recording epochs. Time between epochs and epoch durations may be set independently across acquisition periods according to the study protocol. Triggered epochs may be initiated by the subject using a push-button switch or controlled using an external device. In one of the studies the AZCG system was configured with a 50 Torr pressure-sense switch to allow simultaneous AZCG data acquisition initiated with inflation of an ambulatory blood pressure cuff.

The validity and reliability of that ambulatory impedance cardiograph (AZCG) was tested against the Minnesota Impedance Cardiograph (ZCG) during rest, orthostasis, and mental stress. The devices were compared at two sites in healthy subjects. "In both studies, the AZCG tracked changes across conditions closely with the stationary ICG. Pearson correlation coefficients were 0.65–0.93, and random intraclass correlation coefficients ranged from 0.80 to 0.98, indicating high degrees of shared measurement variance" [14]. Some technical data of that device are presented in Appendices part.

3.8 Other Systems

In this chapter I intend to present other systems that were announced during the scientific conferences.

In Chiang et al. [2] proposed the introduction of a digital signal processing to the recording system, which was the main advantage of their system. They constructed a portable impedance cartography powered by only one 9.6 V Ni/Cd battery that was needed to operate for a one and half hours, continuously. Digital signal processing chip (TMS320C50, National Instrumentation Co.) was used for real-time signal processing, event detection, and SV calculation. Cardiac parameters such as SV, CO, HR and LVET were displayed on a 128×64 dot matrix LCD and updated after each heart beat. This system was also capable of storing the above information in the internal SRAM that can be transmitted to the personal computer through an RS232 interface for further analysis, after the end of monitoring. They used a high frequency constant amplitude current [100 kHz, 4 mA(rms)] is injected into the thorax via application electrodes. Small thoracic impedance changes during each heart beat are detected from receiving electrodes and then amplified, demodulated, filtered and differentiated to extract three signals, dz/dt, Z_0 and ECG, from the thoracic impedance signal. The R peak of the ECG is used as an external interrupt signal by the DSP chip to initiate the event detection and evaluation of SV for each heartbeat.

In 2000, a portable poly-physiograph for non-invasive monitoring of beat-by-beat cardiovascular hemodynamic parameters based on the volume-compensation and electrical-admittance method was described [13]. Their system was able to control measurement procedures, performs blood pressure and CO measurement, automatically processes signals and stores almost 32,000 beats of time-series data. The measurements were based on the volume-compensation method and the transthoracic electrical admittance method. A specialised program for conventional personal computer is used to initialise the measurement system, to extract data from recorder and evaluate the measurement. The device was used for evaluation of a subject's cardiovascular hemodynamic responses to daily physical activities as well as to various psycho-physiological stresses. The measurement system was verified during the subject's cardiovascular hemodynamic responses to daily physical activities as well as to various psycho-physiological stresses.

Twelve subjects were monitored for trials performed in-the-field test for 60–210 min. Less than 3% of the data in each trial were found corrupted by artefacts, which occurred mainly due to body movements.

Another group [15] presented a wearable device able to acquire and to transmit the signal and the parameters extracted for several hours, in order to provide a continuous monitoring. The possibility of wireless Bluetooth® transmission of the signals was the main feature of the system. They considered it, as is useful tool to observe variations in the whole days providing sensible features, making the patient as golden standard of him. This way of investigation would lead to analysis and diagnosis independents from absolute values. The device was composed by a low power sensing and preprocessing board connected through a transmitter to a computer or a PDA. In the sensor part of the system the frequency of the current wave can be chosen in a range between 20 and 100 kHz. They noted that in this way it is possible to avoid the overlapping of other bio-signals, like ECG, or the interference with internal biological processes. In the range they reported the impedance was almost completely resistive and they suggested to neglect the capacitive contribute of the skin and its internal impedance. In order to provide a good accuracy in respect of a variable load, they adopted a specific solution using two transistors (pnp and npn) in a sort of common emitter configuration, ensuring an injected charge equal to zero. The current generator was driven by a low-pass filtered voltage oscillator providing a 32 kHz, sine wave current with 20 µA peak-to-peak amplitude. From the received signal Z_0 was extracted using a low-pass filter ($G = 2$, $f_{\text{flow-pass}} = 1.7$ Hz), and ΔZ by a band-pass filter ($G_{\text{bp}} = 90$, $f_{\text{low-pole}} = 0.08$ Hz, $f_{\text{high-pole}} = 15$ Hz). From ΔZ was obtained the derivative dz/dt.

The authors [15] have seen two possible fields of application of their device: the telemonitoring of a patient to support the clinicians (or a defined and specialized call centre) in a repeated evaluation and observation of the bio-signals of the subject, and in clinical and diagnostic applications through the continuous monitoring of one or more days and then studying the results acquired compared between them to lead to a diagnosis.

References

1. Charloux, A., Lonsdorfer-Wolf, E., Richard, R., Lampert, E., Oswald-Mammosser, M., Mettauer, B., Geny, B., Lonsdorfer, J.: A new impedance cardiograph device for the non-invasive evaluation of cardiac output at rest and during exercise: comparison with the "direct" Fick method. Eur. J. Appl. Physiol. **82**(4), 313–320 (2000)
2. Chiang, C., Hu, W., Shyu, L.: Portable impedance cardiography system for real-time non invasive cardiac output measurement. In: Proceedings, 19th International Conference, IEEE/EMBS, October 30–November 2, pp. 2072–2073 (1997)
3. Cybulski, G.: Computer method for automatic determination of stroke volume using impedance cardiography signals. Acta Physiol. Pol. **39**(5–6), 494–503 (1988)
4. Cybulski, G.: Influence of age on the immediate cardiovascular response to the orthostatic manoeuvre. Eur. J. Appl. Physiol. **73**, 563–572 (1996)
5. Cybulski, G.: Ambulatory impedance cardiography: new possibilities, Letter to the Editor. J. Appl. Physiol. **88**, 1509–1510 (2000)

6. Cybulski, G.: Dynamic impedance cardiography—the system and its applications. Pol. J. Med. Phys. Eng. **11**(3), 127–209 (2005)
7. Cybulski, G., Miśkiewicz, Z., Szulc, J., Torbicki, A., Pasierski, T.: A comparison between impedance cardiography and two dimensional echocardiography methods for measurements of stroke volume (SV) and systolic time intervals (STI). J. Physiol. Pharmacol. **44**(3), 251–258 (1993)
8. Cybulski, G., Książkiewicz, A., Łukasik, W., Niewiadomski, W., Pałko, T.: Ambulatory monitoring device for central hemodynamic and ECG signal recording on PCMCI flash memory cards. In: Computers in Cardiology, pp. 505–507. IEEE, New York, NY, USA (1995)
9. Cybulski, G., Książkiewicz, A., Łukasik, W., Niewiadomski, W., Pałko, T.: Central hemodynamics and ECG ambulatory monitoring device with signals recording on PCMCIA flash memory cards. Med. Biol. Eng. Comput. **34**(suppl 1, part 1), 79–80 (1996)
10. Cybulski, G., Ziółkowska, E., Kodrzycka, A., Niewiadomski, W., Sikora, K., Książkiewicz, A., Łukasik, W., Pałko, T.: Application of impedance cardiography ambulatory monitoring system for analysis of central hemodynamics in healthy man and arrhythmia patients. In: Computers in Cardiology 1997, pp. 509–512. IEEE, New York, NY, USA (1997)
11. Cybulski, G., Ziółkowska, E., Książkiewicz, A., Łukasik, W., Niewiadomski, W., Kodrzycka, A., Pałko, T.: Application of impedance cardiography ambulatory monitoring device for analysis of central hemodynamics variability in atrial fibrillation. In: Computers in Cardiology, vol. 26 (Cat. No. 99CH37004), pp. 563–566. IEEE, 1999, Piscataway, NJ, USA (1999)
12. McFetridge-Durdle, J.A., Routledge, F.S., Parry, M.J., Dean, C.R., Tucker, B.: Ambulatory impedance cardiography in hypertension: a validation study. Eur. J. Cardiovasc. Nurs. **7**(3), 204–213 (2008)
13. Nakagawara, M., Yamakoshi, K.: A portable instrument for non-invasive monitoring of beat-by-beat cardiovascular haemodynamic parameters based on the volume-compensation and electrical-admittance method. Med. Biol. Eng. Comput. **38**(1), 17–25 (2000)
14. Nakonezny, P.A., Kowalewski, R.B., Ernst, J.M., Hawkley, L.C., Lozano, D.L., Litvack, D.A., Berntson, G.G., Sollers 3rd, J.J., Kizakevich, P., Cacioppo, J.T., Lovallo, W.R.: New ambulatory impedance cardiograph validated against the Minnesota impedance cardiograph. Psychophysiology **38**(3), 465–473 (2001)
15. Panfili, G., Piccini, L., Maggi, L., Parini, S., Andreoni, G.: A wearable device for continuous monitoring of heart mechanical function based on impedance cardiography. Conf. Proc. IEEE Eng. Med. Biol. Soc. **1**, 5968–5971 (2006)
16. Parry, M.J., McFetridge-Durdle, J.: Ambulatory impedance cardiography: a systematic review. Nurs. Res. **55**(4), 283–291 (2006)
17. Qu, M., Webster, J.G., Tompkins, W.J., Voss, S., Bogenhagen, B., Nagel, F.: Portable impedance cardiograph for ambulatory subjects. In: Proceedings of the Ninth Annual Conference of the IEEE Engineering in Medicine and Biology Society, vol. 3, pp. 1488–1489. IEEE, New York, NY, USA (1987)
18. Riese, H., Groot, P.F., van den Berg, M., Kupper, N.H., Magnee, E.H., Rohaan, E.J., Vrijkotte, T.G., Willemsen, G., de Geus, E.J.: Large-scale ensemble averaging of ambulatory impedance cardiograms. Behav. Res. Methods Instrum. Comput. **35**(3), 467–477 (2003)
19. Sherwood, A., McFetridge, J., Hutcheson, J.S.: Ambulatory impedance cardiography: a feasibility study. J. Appl. Physiol. **85**(6), 2365–2369 (1998)
20. Willemsen, G.H., De Geus, E.J., Klaver, C.H., Van Doornen, L.J., Carroll, D.: Ambulatory monitoring of the impedance cardiogram. Psychophysiology **33**(2), 184–193 (1996)
21. Zhang, Y., Qu, M., Webster, J.G., Tompkins, W.J.: Impedance cardiography for ambulatory subjects. In: Proceedings of the Seventh Annual Conference of the IEEE/Engineering in Medicine and Biology Society. Frontiers of Engineering and Computing in Health Care, vol. 2, pp. 764–769. IEEE, New York, NY, USA (1985)

Chapter 4
Validation of the Ambulatory Impedance Cardiography Method

In this chapter it is described the problem of ambulatory impedance cardiography validation against the clinically accepted reference methods, basing on data from own research. It also contains the discussion of the motion artefacts in ambulatory impedance cardiography.

4.1 Introduction

There are two major questions regarding reliability of ambulatory ICG. First of all it is a problem of accuracy of ICG and its validation using a clinically accepted reference method. Although the ICG was compared with all invasive and non-invasive methods, its precision is still discussed [9, 17]. This uncertainty is directly transferred to the ambulatory version of the method. Moreover, the comparisons were performed, mainly, in a supine, whereas ambulatory ICG should be also verified in a vertical position, the natural position of patient during a large part of holter-type recordings. This is even more important, since some authors suggested the hypothesis that SV measured by ICG in standing position gives underestimated values [38]. Second one is a problem of high sensitivity of the ICG signals to the motion artefacts. This sensitivity was observed during exercise tests performed in the laboratory, which are only a simulation of the "real life" movement of the patient during holter-type recordings. Since it is not possible to remove the motion artefacts, it is essential to know the rate of artefacts during standard impedance holter recordings. This would help to evaluate how representative are the mean values from periods of different length for evaluating trends.

Since my intention was to verify the ambulatory ICG method let me present the results of two studies regarding the comparison between ICG and pulsed Doppler measurement in supine and tilted positions, and the estimation of artefact rate in ambulatory ICG examination in healthy men and patients with atrial fibrillation.

G. Cybulski, *Ambulatory Impedance Cardiography*,
Lecture Notes in Electrical Engineering, 76, DOI: 10.1007/978-3-642-11987-3_4,
© Springer-Verlag Berlin Heidelberg 2011

Both studies were performed using ReoMonitor. The following parts are based on the results published earlier in Cybulski et al. [13, 15], Cybulski [14].

4.2 Validation using Reference Methods

4.2.1 Background and Motivation

It would appear that the problem of obtaining non-invasive, continuous and accurate measurement of stroke volume (SV) in various, clinical and physiological, situations could be solved by applying impedance cardiography as a safe, simple and inexpensive method [38]. However, the precision and accuracy of this method are still subject to discussion [9, 17]. There have been several papers showing the comparison of the results of SV measurements by ICG with those obtained by other, invasive and non-invasive, methods. Some authors accepted ICG as a reliable method for determining both absolute SV values and changes in them [2, 19, 28, 29, 33, 39]. However other authors [5, 23, 32], while accepting the validity of ICG for measuring changes in SV, have expressed reservations about applying this method to calculations of the absolute values of SV and have suggested a need for further methodical investigations. [2], compared SV determined by M-mode echocardiography with the SV values simultaneously measured by ICG and concluded that the ICG method should be used for evaluation of the trends in SV responses to physiological or pharmacological interventions rather than for estimation of absolute values. Antonicelli et al. [1] concluded that ICG might represent a reliable alternative to pulsed Doppler echocardiography for the non-invasive estimation of cardiac output at rest in elderly patients. Scherhag et al. [39] performed a comparison between ICG and stress echocardiography results after administration of dobutamine and dipyridamole. They concluded that automated ICG not only allows surveying and monitoring hemodynamic changes during pharmacological stress echocardiography but also contributes to differentiation of pathologic stress responses.

In several studies ICG has been validated simultaneously using the direct Fick [8] and the CO_2 re-breathing method [9, 36]. Also, Drazner et al. [17], comparing ICG with two invasive methods, thermodilution and Fick, in patients with heart failure secondary to ischemic or non-ischemic cardiomiopathy, found that invasive measures of cardiac output were significantly correlated with ICG. Their opinion was not supported by Leslie et al. [30], who compared impedance cardiography (thoracic bioimpedance) with thermodilution in 11 patients with stable chronic heart failure. Their study demonstrated a correlation between both techniques but shows a poor level of agreement. The impedance method underestimated cardiac output compared with thermodilution, and this difference appeared greater with higher cardiac outputs. Agreement was worse when results were expressed as changes from baseline. Their study does not support the use of thoracic bioimpedance in its current form as an alternative to thermodilution in stable patients with chronic heart failure.

In the earlier comparison [11] the correlation coefficient ($r = 0.69$) was obtained between SV calculated using ICG and pulsed Doppler echocardiography in an apex approach. The subjects, due to certain technical requirements, remained in a supine position and there were no beat-to-beat comparison data from a vertical position.

The aim of this comparison was to validate the ICG data [SV, ejection time (ET), and pre-ejection period (PEP)] obtained applying the ReoMonitor [11, 12] central hemodynamics ambulatory monitoring device, in two positions using a commonly accepted, non-invasive reference method, e.g. pulsed Doppler echocardiography.

The comparison was performed during a tilt test, since the analysis of some hemodynamic parameters [SV, and systolic time intervals (STI)] obtained during postural tests could help in early detection of the mechanism of orthostatic syndrome [4]. Also the significance of postural tests is growing in clinical practice [7, 25]. Moreover, verification of data from the vertical position would enable reliable evaluation of long time recording obtained with ambulatory ICG during normal daytime activity in both healthy subjects and patients (e.g. arrhythmia, pharmacological studies).

4.2.2 Experimental Studies

The examinations were carried out with 13 young healthy volunteers (six men and seven women aged 23–33 years), who gave their written informed consent. The subjects remained recumbent for at least 15 min before examination and during the measurement with their heads elevated slightly above the level of their legs. The hemodynamic response was monitored continuously by ambulatory impedance cardiography. The echocardiographic acquisitions were performed two times: 1 min before and 10 min after a 60° head-up tilting manoeuvre. For each subject in each position from 14 to 25 cycles were recorded using echocardiography.

A Sonos 5500 ultrasound imaging system with two-dimensional, M-mode, continuous and pulsed Doppler facilities, was used for measuring SV in the ascending aorta using a suprasternal projection. SV was calculated as a product of flow velocity integral (FVI) and the area of aorta cross section (ACS) [21]. After mapping the ascending aorta to obtain the highest velocities with minimal dispersion of Doppler signal and angle correction by 2-dimensional colour Doppler study, pulsed Doppler aortic flow velocities were recorded. The FVI was determined automatically after digitising the brightest part of the spectral display, using technically acceptable spectral displays with a minimal dispersion of the Doppler signal at high velocities [18]. The ACS was estimated using two dimensional (2D-mode) echocardiography techniques in parasternal long axis images. The smallest systolic diameter of the ascending aorta from inner to inner wall was measured perpendicularly to the aortic lumen superior to the Valsalva sinus and averaged from 3 to 4 single measurements during the left ventricular ejection period before measurement of FVI [21]. The pre-ejection period was measured as the time between the ECG Q-wave and the beginning of aortic flow. The ejection time was measured between the beginning and the end of the aortic flow [31].

Simultaneously, the ECG (2nd lead) and ICG signals from ReoMonitor (impedance cardiography ambulatory monitoring device) were recorded on PCMCIA Flash Cards. Chapter 3 above contains a detailed description of the system; some research applications have been published elsewhere [12, 26, 27, 40, 42].

Stroke volume was calculated according to the Kubicek formula, assuming constant blood resistivity $\rho = 130$ Ωcm.

The relationships between the values measured by these methods were evaluated by calculation of the correlation coefficient and performance of a linear regression analysis. Additionally, the t Student test for paired values was used to evaluate the differences between the simultaneously obtained ICG and echocardiography results of SV. For hypothesis testing the $p < 0.05$ level was used. The bias was expressed at the level of 95% confidence, as a mean difference between measured values (SV, ET, PEP) $\pm 2SD$.

4.2.3 Results of the Own Experimental Studies

Figure 4.1 consists of the correlation plot for stroke volume (SV), ejection time (ET), and pre-ejection period (PEP) between the values obtained simultaneously by automatised impedance cardiography (SV_{ic}, ET_{ic}, PEP_{ic}) and pulsed wave Doppler echocardiography (SV_{echo}, ET_{echo}, PEP_{echo}) in 13 young, healthy subjects. The bias plot of differences between the values obtained by two methods against the echo-graphically measured values is presented in Fig. 4.2.

Stroke Volume

Stroke volume measured with echocardiography (SV_{echo}) was similar to stroke volume measured with impedance cardiography (SV_{icg}) in a supine position (71.2 $\pm$ 12.3 vs. 72.3 $\pm$ 13.3 ml, NS), and slightly different in a tilted one— (38.1 $\pm$ 8.1 vs. 43.9 $\pm$ 10.3 ml, $d = -5.8 \pm 8.2$ ml, $p < 0.001$). The difference $SV_{icg} - SV_{echo}$ (bias) measured for both positions (expressed as mean $\pm$ 2SD) was equal to -3.6 ± 19.2 ml. Individual correlation coefficients for SV varied within a range of 0.75–0.93. For the minimal value of r the slope (b) was 0.483 and the intercept (a) was 29.6. For maximal value of r the regression parameters were $b = 0.798$ and $a = 15.0$. The minimal and maximal individual slopes (and the corresponding intercepts) were 0.483 (29.6) and 1.17 (−3.0), respectively. The linear regression between measured values obtained for all subjects was described using the formula: $SV_{icg} = 13.9 + 0.813 * SV_{echo}$ ($r = 0.857$, SEE = 9.03, $n = 496$).

Fig. 4.1 The correlation plot for stroke volume (SV), ejection time (ET) and pre-ejection period (PEP) between the values obtained simultaneously by automatized impedance cardiography (ic-indexed) and pulsed wave Doppler echocardiography (echo-indexed) in 13 young, healthy subjects. (Adapted from Cybulski et al. [13], and Cybulski [14])

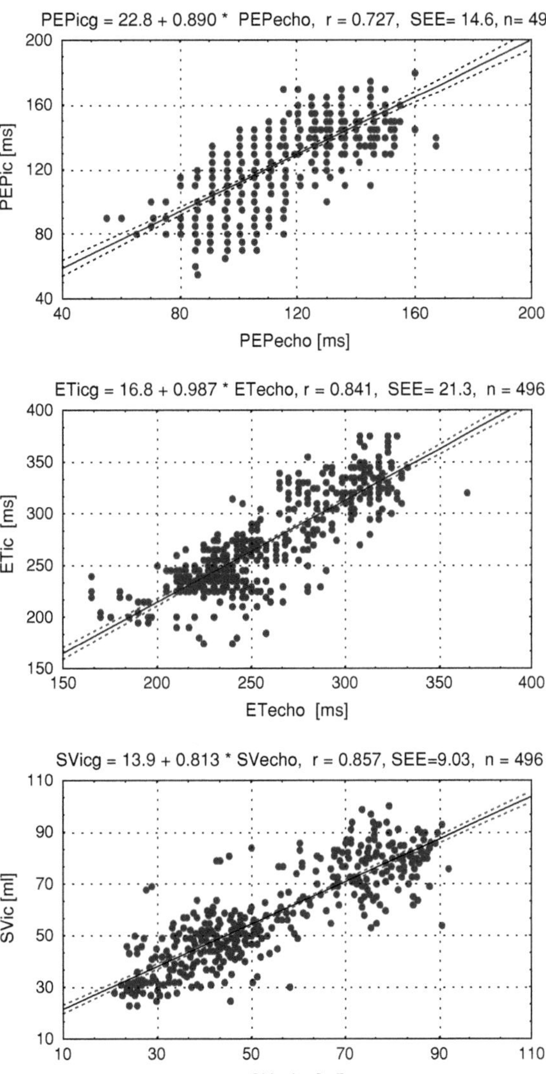

Ejection Time

The ejection time measured by echocardiography (ET_{echo}) was significantly lower than the ejection time measured by impedance cardiography (ET_{icg}) in both positions: supine (294.4 ± 26.7 vs. 310.5 ± 26.8 ms, $d = -16.1 ± 19.8$ ms, $p < 0.001$), and tilted—(233.3 ± 24.9 vs. 239.4 ± 24.6 ms, $d = -6.0 ± 24.5$ ms, $p < 0.002$).

The level of this difference was higher in the supine position (<5.6% of the measured value) than in the tilted one (<2.7% of the measured value).

Fig. 4.2 The bias plot of the differences between the values obtained by two methods against the echographically measured stroke volume (SV), ejection time (ET) and pre-ejection period (PEP). (Adapted from Cybulski et al. [13], and Cybulski [14])

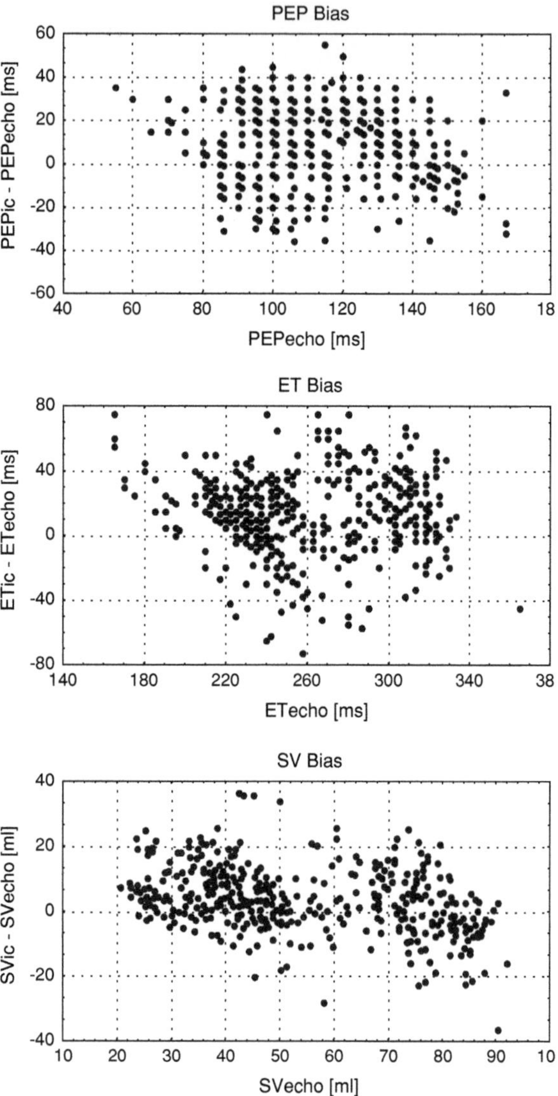

The ET bias (difference $ET_{icg} - ET_{echo}$, expressed as mean $\pm$ 2SD) was equal to -13.4 ± 50 ms. Individual correlation coefficients for ET varied within a range of 0.65–0.94. For the minimal value of r, the slope b was 0.893 and the intercept was 37. For maximal value of r, the regression parameters were $b = 1.011$ and $a = 1$. The minimal and maximal individual slopes (and the corresponding intercepts) were 0.746 (83) and 1.342 (-87), respectively. The linear regression between measured values obtained for all subjects was described using the formula: $ET_{icg} = 16.8 + 0.987 * ET_{echo}$ ($r = 0.841$, SEE $= 21.3$, $n = 496$).

Pre-ejection Period

The pre-ejection period with echocardiography (PEP_{echo}) was significantly lower than the pre-ejection period measured with impedance cardiography (PEP_{icg}) in both positions, supine (97.2 ± 13.0 vs. 103.7 ± 21.1 ms, $d = -6.5 \pm 19.1$ ms, $p < 0.001$), and tilted—(127.8 ± 15.7 vs. 140.4 ± 14.6 ms, $d = -12.6 \pm 16.1$ ms, $p < 0.001$). The level of this difference was lower in the supine position ($<6.3\%$ of the measured value) than in the tilted position ($<9.9\%$ of the measured value). The PEP bias (difference $PEP_{icg} - PEP_{echo}$, expressed as mean $\pm 2SD$) was equal to -10.3 ± 36 ms. Individual correlation coefficients for PEP varied within a range of 0.52–0.94. For the minimal value of r, the b was 0.911 and $a = 29.9$. For maximal value of r the regression parameters were $b = 1.108$ and $a = 14.9$. The minimal and maximal individual slopes (and the corresponding intercepts) were 0.473 (70.7) and 1.432 (-36.9), respectively. The linear regression between measured values obtained for all subjects was described using the formula: $PEP_{icg} = 22.8 + 0.890^* PEP_{echo}$ ($r = 0.727$, SEE $= 14.6$, $n = 496$).

4.2.4 Discussion and Conclusions

ICG is an empirical method and it remains unclear how many and what factors influence an ICG signal. So it is far from obvious that SV values measured with this method during lying and standing/tilt are directly comparable. It cannot be excluded that SV in standing/tilt is systematically biased due to factors other than SV, which have been modified by the change of body position. This problem should be addressed if the potential of ICG is to be fully exploited.

There are several potential sources of discrepancy between the measurements performed by ICG and echocardiography:

1. errors in ACS measurement which disturb the results in the second power,
2. translocation of the sample volume in relation to the aortic valve [16],
3. disturbances in the aortic velocity profile assumed laminar [37],
4. differences in blood resistivity, which has been assumed to be constant [22],
5. the methodical inaccuracy of automatizing the ICG measurement (particularly in determining the onset of ejection which affects both PEP and ET measurements) [10].

In this study it has been possible only to estimate the compound effect of superposition of these errors.

In the supine position there were no significant differences between SV values obtained using the two methods. The variance of echocardiography SV measurements was slightly lower than the ICG equivalent. Mean SV measured with ICG and echocardiography attains similar values, with the tendency of SV_{ic} being about 6 ml greater than SV_{echo}. This tendency was observed only in tilt. This observation is, however, in contrary to the generally held hypothesis that SV measured by ICG in standing position gives underestimated values. It has been claimed that the huge impact (inverted second order) of the increased Z_0 (in standing position) according to the Kubicek formula causes this underestimation. In fact it cannot be excluded that the reference method, gives underestimated results in tilt.

Considering the level of variations in echocardiography and impedance cardiography measurements caused by physical changes as well as the component of variance due to reproducibility error [20] the value of $r = 0.857$, for this comparison, seems to be close to the achievable clinical correlation coefficient.

Papers have not been found in the relevant literature that questions the accuracy of STI absolute values measurement by ICG. Although the differences between ET (and PEP) measured by both methods in both situations are significant, the level of discrepancy does not exceed 5.6% (9.9%) of the measured value. Also, the variances of ET are similar. The variance of PEP measured by ICG in supine position is higher than measured by echo or ICG in tilt. The potential source of error in systolic time interval measurement by ICG is the uncertainty of proper determination of the ejection onset. Also, the spatially averaged ICG signal (including the part of the neck arteries) could be slightly delayed (10–20 ms) in comparison to the signal obtained directly from the aortic valve. This could cause the higher values of PEP by ICG (the same beginning at Q point of QRS complex and delayed ending point). The level of discrepancy between ICG and echocardiography values of STI seems to be acceptable, particularly for ET.

The results justified the use of ICG to differentiate between mean values from the groups of subjects. The change of SV due to body position alteration is significant and absolute values are similar in two methods. This supports long held opinion that ICG measurement renders itself to reliable estimation of SV change. On the basis of these results it may be concluded that the mean SV change in a group of subjects caused by the change of body position can be reliably measured using an ambulatory version of the ICG method. Thus ambulatory impedance cardiography appears to be a useful method for analysis of hemodynamic changes during postural tests.

4.3 The Quality of the Ambulatory Impedance Cardiography Recordings

4.3.1 Background and Motivation

It is well known among ICG users that this method is very sensitive to motion artefacts. Although it is relatively simple task to perform the automatic analysis of the clear ICG signals, the noisy signals analysis sometimes seems be not possible. Figure 4.3 presents the example of the noisy ECG and ICG ambulatory recordings that could give the impression of the level of difficulties for automatic analysis

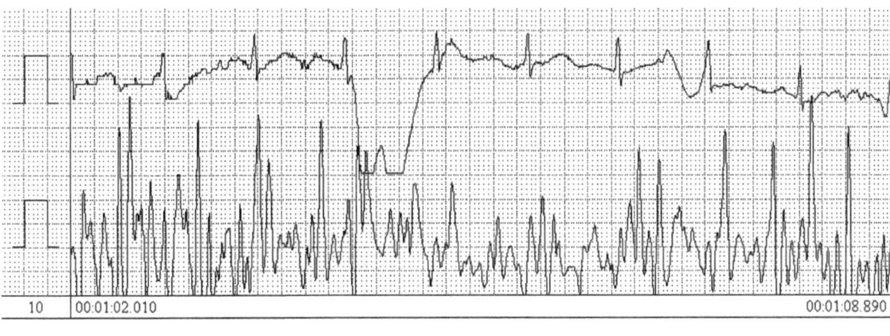

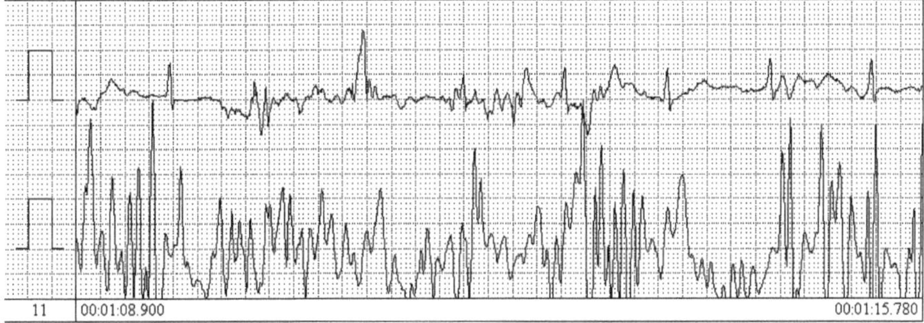

Fig. 4.3   The noisy ECG and ICG ambulatory recordings

program designers. Despite the signal analysis efforts when motion creates a loss of the useful signal, even the most sophisticated program is useless. This points out how the careful placement of electrodes, their positioning and firm fixing is important, since these endeavours can significantly reduce the rate of artefacts [6]. A fundamental question regarding ICG ambulatory monitoring was evaluation of the rate of artefacts during standard impedance Holter recordings. It is also important to know what is the average length of recordings, which are too noisy to be analysed. Another problem is the question of how representative are the mean values from periods of different length for evaluating trends.

4.3.2 Experimental Studies

12 male patients with atrial fibrillation aged 36–78 years and 6 healthy males (24–68 years) as a control underwent examination lasting for 12 h starting from about 18:00.

The ratio of artefact free cycles to the total number of cycles was calculated for the same hour of recording (during daytime activity) when the averaging period varied from 5 to 60 min (R05, R10, …, R60). The maximal and minimal values of Rxx were found for each subject in every averaging period.

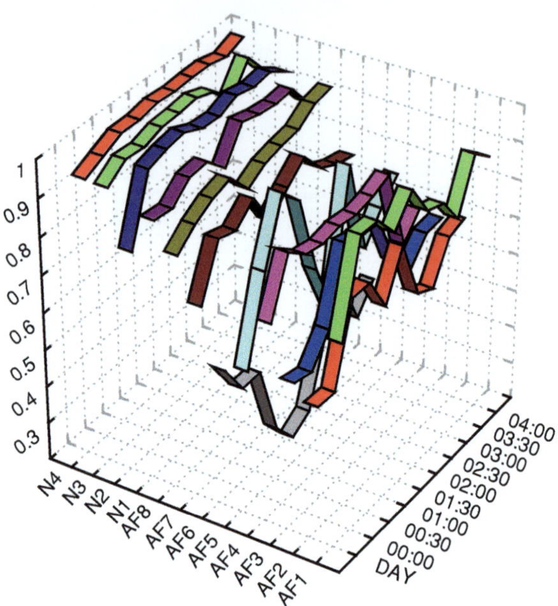

Fig. 4.4 The artefact free to total number of cycles ratio during daytime activity and the night hours obtained within 30 min intervals in 8 patients with atrial fibrillation (AF1···AF8) and 4 healthy subjects (N1···N4). (Adapted from Cybulski et al. [15])

Also the rates of artefact-free recordings were calculated for each 30-min period during the daytime (18.00–23.00) and over the night period (23:00–04:00).

4.3.3 Results of the Experimental Studies

The percentage of cycles recognised as normal according to the criteria of automatic determination of cardiac parameters from impedance cardiography signals varied from 20% during the daytime up to 90% during the night hours, when 5-min period are analysed. However, this index calculated for each 30 min was always higher than 60% in both patients and healthy subjects.

Figure 4.3 presents the artefact-free to total number of cycles ratio during daytime activity and the night hours obtained within 30 min intervals for 8 patients with atrial fibrillation and 4 healthy subjects.

Figure 4.4 presents the values calculated as the mean ± SD of the individual maximal and minimal ratios (artefact-free to all cycles) obtained for each subject. Individual maximal and minimal ratios were taken from each of the following periods: 5, 10, 15, 20, 30, and 60 min, named R05 to R60, respectively. Data were analysed for the same hour of the recording in all subjects. In the worst case, the minimal individual value of the ratio (observed in 5-min periods) was 0.21 (when mean ± SD was 0.63 ± 0.26, range 0.21–0.98). In the recordings of the best quality the minimal value of the ratio was 0.46 (when mean ± SD was 0.83 ± 0.17, and range 0.46–0.99). For the entire group of subjects ($n = 18$) in

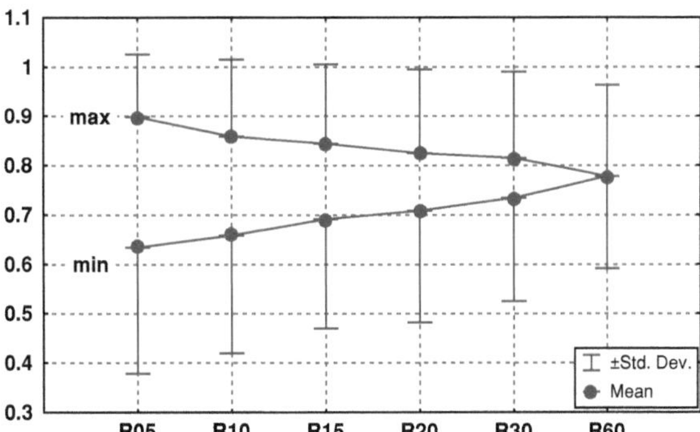

Fig. 4.5 The artefact free to all cycles ratios, presents as the mean ± SD of the individual maximal and minimal ratios obtained for each subject. Individual max and min ratios were calculated for each of the following periods: 5, 10, 15, 20, 30, and 60 min, named R05 to R60, respectively. (Adapted from [14])

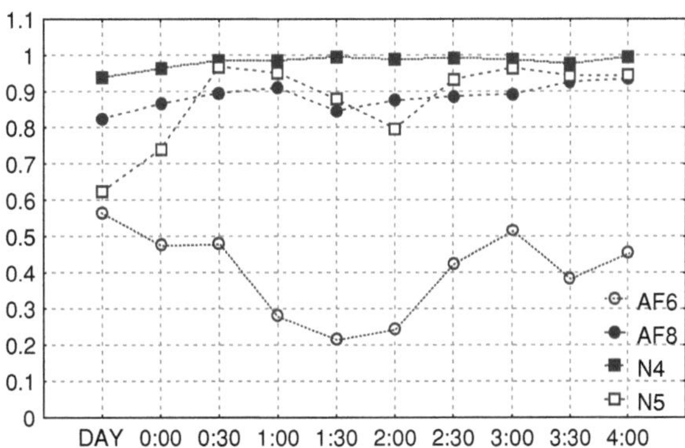

Fig. 4.6 The artefact free to total number of cycles ratio during day-time activity and the night hours obtained within 30 min intervals for 2 patients with atrial fibrillation (AF6, AF8) and 2 healthy subjects (N4, N5). (Adapted from Cybulski et al. [15])

two cases the minimal ratio was lower than 0.25 and in eight cases more than 0.75 (Fig. 4.5).

Figure 4.6 presents the artefact-free to total number of cycles ratio during daytime activity and the night hours obtained within 30 min intervals for 2 patients with atrial fibrillation and 2 healthy subjects. The recordings obtained for the patient AF6 and AF8 were of the poorest and the best quality, respectively. In healthy subjects the poorest quality was observed in N5 and the best one in N4.

Figure 4.6 the artefact free to total number of cycles ratio during day-time activity and the night hours obtained within 30 min intervals for 2 patients with atrial fibrillation (AF6, AF8) and 2 healthy subjects (N4, N5). (Adapted from Cybulski et al. [15]).

4.3.4 Discussion and Conclusions

The percentage of cycles recognised as normal according to the criteria of automatic determination of cardiac parameters from impedance cardiography signals varied from 20% during the daytime up to 90% during the night hours within

5-min periods. It was observed that speaking and vigorous movements distinctly decrease the number of artefact free cycles. For night recordings the percentage of cycles recognised as normal achieved the level of 75–90%. The rate of the artefact-free cycles was markedly lower during the normal daytime activity than for nighttime recordings. However, it seems that during cardiac events (AF, tachycardia, etc.), as well as after them (and sometimes before them) the subject is relatively less active physically in comparison to the periods without events [34]. Thus, daytime recordings (higher level of artefacts) seem to be justified, because it is possible to find a sufficient number of artefact-free cycles over the period of interest. This is in accordance with the findings of [3] who used an AIM-8-V3 Wearable Cardiac Performance Monitor (Bio-impedance Technology, Inc., USA) to study the reproducibility of daytime and nighttime ambulatory bioimpedance-derived measures of hemodynamic function in young men. They reported that across 2 months daytime and night-time ambulatory bioimpedance-derived measures of HR and heather index (HI) in young men were highly repeatable and SV, CO, PEP and LVET were moderately repeatable. They also suggested that ambulatory ICG methodology should prove useful in cardiovascular research and clinical care.

However, Willemsen et al. [41] when verifying their system (the VU-AMD ambulatory monitor for impedance cardiography) concluded that it is a valid device for the measurement of systolic time intervals in real-life situations but that its applicability for absolute stroke volume and cardiac output determination remained to be established. Their reservations regarded, however, stroke volume and cardiac output values only during exercise. Nakonezny et al. [35], analysed the validity and reliability of the ambulatory impedance cardiograph (AZCG) against the Minnesota Impedance Cardiograph (ZCG) during rest, orthostasis, and mental stress. They performed the comparison because reliable ambulatory device would allow studies outside the lab. The devices were compared at two sites on healthy subjects. In both studies, the AZCG tracked changes across conditions closely with the ZCG (all period vs. device interactions were insignificant). Pearson correlation coefficients were 0.65 to 0.93, random intraclass correlation coefficients ranged from 0.80 to 0.98, indicating high degrees of shared measurement variance, and Cronbach's alpha coefficient of reliability indicated very good internal reliabilities (0.91 to 0.99). They concluded that AZCG appeared to provide valid and reliable estimates of cardiac function at rest and during behavioural challenges in the lab.

Kelsey et al. [24] analysed impedance cardiograms during psychological tests using the following methods: a conventional method, involving ensemble averaging after careful editing of beat-to-beat waveforms, and a streamlined method, involving ensemble averaging without beat-to-beat editing. The side effect of their work is the conclusion that "variations in beat-to-beat editing do not constitute a serious source of error in the ensemble-averaged impedance cardiogram". This supports the opinion that ICG automatic analysis without manual data editing could give the same physiological response to the test (if there is any) as careful (and slow) manual interpretation.

In conclusion, body movements and speaking significantly distort the ICG signal, which reduces the quantity of useful data. However, even when a large number of artefacts occur (during in practice a few short periods) recordings still make it possible to perform automatic evaluation of cardiac parameters. Also there is easy access to single beat data and the synthetic index characterises the quality of the signal and its reliability. It was demonstrated that the system might be used to collect signals in a laboratory and in the field to monitor changes in cardio-vascular parameters.

References

1. Antonicelli, R., Savonitto, S., Gambini, C., Tomassini, P.F., Sardina, M., Paciaroni, E.: Impedance cardiography for repeated determination of stroke volume in elderly hypertensives: correlation with pulsed Doppler echocardiography. Angiology 42(8), 648–653 (1991)
2. Aust, P.E., Belz, G.G., Belz, G., Koch, W.: Comparison of impedance cardiography for measurement of stroke volume. Eur. J. Clin. Pharmacol. 23(6), 475–477 (1982)
3. Barnes, V.A., Johnson, M.H., Treiber, F.: Temporal stability of twenty-four-hour ambulatory hemodynamic bioimpedance measures in African American adolescents. Blood Pressure Monit. 9(4), 173–177 (2004)
4. Bellard, E., Fortrat, J.O., Schang, D., Dupuis, J.M., Victor, J., Leftheriotis, G.: Changes in the transthoracic impedance signal predict the outcome of a 70 degrees head-up tilt test. Clin. Sci. (Lond) 104, 119–126 (2003)
5. Boer, P., Roos, J.C., Geyskes, G.G., Mees, E.J.D.: Measurement of cardiac output by impedance cardiography under various conditions. Am. J. Physiol. 237(4), 491–492 (1979)
6. Bogaard, H.J., Woltjer, H.H., Postmus, P.E., de Vries, P.M.: Assessment of the haemodynamic response to exercise by means of electrical impedance cardiography: method, validation and clinical applications. Physiol. Meas. 18(2), H95–H105 (1997)
7. Brignole, M., Alboni, P., Benditt, D., Bergfeldt, L., Blanc, J.J., Bloch Thomsen, P.E., van Dijk, J.G., Fitzpatrick, A., Hohnloser, S., Janousek, J., Kapoor, W., Kenny, R.A., Kulakowski, P., Moya, A., Raviele, A., Sutton, R., Theodorakis, G., Wieling, W.: Task force on syncope, European Society of Cardiology. Guidelines on management (diagnosis and treatment) of syncope. Eur. Heart J. 22, 1256–1306 (2001)
8. Charloux, A., Lonsdorfer-Wolf, E., Richard, R., Lampert, E., Oswald-Mammosser, M., Mettauer, B., Geny, B., Lonsdorfer, J.: A new impedance cardiograph device for the non-invasive evaluation of cardiac output at rest and during exercise: comparison with the "direct" Fick method. Eur. J. Appl. Physiol. 82(4), 313–320 (2000)
9. Christensen, T.B., Jensen, B.V., Hjerpe, J., Kanstrup, I.L.: Cardiac output measured by electric bioimpedance compared with the CO_2 rebreathing technique at different exercise levels. Clin. Physiol. 20(2), 101–105 (2000)
10. Cybulski, G.: Computer method for automatic determination of stroke volume using impedance cardiography signals. Acta Physiol. Pol. 39(5–6), 494–503 (1988)
11. Cybulski, G., Szulc, M.J., Torbicki, A., Pasierski, T.: A comparison between impedance cardiography and two dimensional echocardiography methods for measurements of stroke volume (SV) and systolic time intervals (STI). J. Physiol. Pharmacol. 44(3), 251–258 (1993)
12. Cybulski, G., Ziółkowska, E., Książkiewicz, A., Łukasik, W., Niewiadomski, W., Kodrzycka, A., Pałko, T.: Application of impedance cardiography ambulatory monitoring device for analysis of central hemodynamics variability in atrial fibrillation. In: Computers in Cardiology, vol. 26 (Cat. No.99CH37004). IEEE, pp. 563–566. Piscataway, NJ, USA (1999)

13. Cybulski, G., Michalak, E., Koźluk, E., Piątkowska, A., Niewiadomski, W.: Stroke volume and systolic time intervals: beat-to-beat comparison between echocardiography and ambulatory impedance cardiography in supine and tilted positions. Med. Biol. Eng. Comput. **42**, 707–711 (2004)

14. Cybulski, G.: Dynamic impedance cardiography—the system and its applications. Pol. J. Med. Phys. Eng. **11**(3), 127–209 (2005)

15. Cybulski, G., Niewiadomski, W., Gasiorowska, A., Kwiatkowska, D.: Signal quality evaluation in ambulatory impedance cardiography. In: IFMBE Proceedings vol. 17 (Ed. Scharfetter, Merwa), 13th International Conference on Electrical Bioimpedance, Graz, Austria, Aug 29th–Sept 2nd, pp. 590–559 (2007)

16. Dittmann, H., Voelker, W., Karsch, K.R., Seipel, L.: Influence of sampling site and flow area on cardiac output measurement by Doppler echocardiography. J. Am. Coll. Cardiol. **10**, 818–823 (1987)

17. Drazner, M.H., Thompson, B., Rosenberg, P.B., Kaiser, P.A., Boehrer, J.D., Baldwin, B.J., Dries, D.L., Yancy, C.W.: Comparison of impedance cardiography with invasive hemodynamic measurements in patients with heart failure secondary to ischemic or nonischemic cardiomyopathy. Am. J. Cardiol. **89**(8), 993–995 (2002)

18. Dubin, J., Wallerson, D.C., Cody, R.J., Devereux, R.B.: Comparative accuracy of Doppler echocardiographic methods for clinical stroke volume determination. Am. Heart J. **120**, 116–123 (1990)

19. Ebert, T.J., Eckberg, D.L., Vetrovec, G.M., Cowley, M.J.: Impedance cardiograms reliably estimate beat-by-beat changes of left ventricular stroke volume in humans. Cardiovasc. Res. **18**, 354–360 (1984)

20. Francis, D.P., Coats, A.J.S., Gibson, D.G.: How high can correlation coefficient be? Effect of limited reproducibility on common cardiological variables. Int. J. Cardiol. **69**, 185–199 (1999)

21. Gardin, J.M., Tobis, J.M., Dabestani, A., Smith, C., Elkayam, U., Castleman, E., White, D., Allfie, A., Henry, W.L.: Superiority of two-dimensional measurements of aortic vessel diameter in Doppler echocardiographic estimates of left ventricular stroke volume. J. Am. Coll. Cardiol. **6**, 66–74 (1985)

22. Geddes, L.A., Sadler, C.: The specific resistance of blood at body temperature. Med. Biol. Eng. **5**, 336–339 (1973)

23. Judy, W.V., Langley, F.M., McCowen, K.D., Stinned, D.M., Backer, L.E., Johnson, P.C.: Comparative evaluation of the thoracic impedance and isotope dilution methods for measuring cardiac output. Aerospace Med. **40**(5), 532–536 (1969)

24. Kelsey, R.M., Reiff, S., Wiens, S., Schneider, T.R., Mezzacappa, E.S., Guethlein, W.: The ensemble-averaged impedance cardiogram: an evaluation of scoring methods and interrater reliability. Psychophysiology **35**(3), 337–340 (1998)

25. Kenny, R.A., Ingram, A., Bayliss, J., Sutton, R.: Head-up tilt: a useful test for investigating unexplained syncope. Lancet **2**, 1352–1354 (1986)

26. Koźluk, E., Piątkowska, A., Wołkanin-Bartnik i wsp, J.: Permanent AAI pacing in patient with vasovagal syndrome—case study (in Polish). Folia Cardiol. **8**, 28 (2001)

27. Koźluk, E., Cybulski, G., Szufladowicz i wsp, E.: Application of Reomonitor in optimizing the pacemaker with rate drop response (in Polish). Folia Cardiol. **8**, 42 (2001)

28. Kubicek, W.G., Karnegis, J.N., Patterson, R.P., Witsoe, D.A., Mattson, R.H.: Development and evaluation of an impedance cardiac output system. Aerospace Med. **37**(12), 1208–1212 (1966)

29. Kubicek, W.G., Patterson, R.P., Witsoe, D.A.: Impedance cardiography as a non-invasive method for monitoring cardiac function and other parameters of the cardiovascular system. Ann. N.Y. Acad. Sci. **170**, 724–732 (1970)

30. Stephen, J.L., Sinead, M., David, E.N., David, J.W., Martin, A.D.: Non-invasive measurement of cardiac output in patients with chronic heart failure. Blood Pressure Monit. **9**(5), 277–280 (2004)

31. Loutfi, H., Nishimura, R.A.: Quantitative evaluation of left ventricular systolic function by Doppler echocardiographic techniques. Echocardiography **11**, 305–314 (1994)
32. Milsom, I., Sivertsson, R., Biber, B., Olsson, T.: Measurement of stroke volume with impedance cardiography. Clin. Physiol. **2**, 409–417 (1982)
33. Muzzi, M., Jeutter, D.C., Smith, J.J.: Computer-automated impedance-derived cardiac indexes. IEEE Trans. Biomed. Eng. **33**(1), 42–47 (1986)
34. Nakagawara, M., Yamakoshi, K.: A portable instrument for non-invasive monitoring of beat-by-beat cardiovascular haemodynamic parameters based on the volume-compensation and electrical-admittance method. Med. Biol. Eng. Comput. **38**(1), 17–25 (2000)
35. Nakonezny, P.A., Kowalewski, R.B., Ernst, J.M., Hawkley, L.C., Lozano, D.L., Litvack, D.A., Berntson, G.G., Sollers 3rd, J.J., Kizakevich, P., Cacioppo, J.T., Lovallo, W.R.: New ambulatory impedance cardiograph validated against the Minnesota Impedance Cardiograph. Psychophysiology **38**(3), 465–473 (2001)
36. Niizeki, K., Miyamoto, Y., Doi, K.: A comparison between cardiac output determined by impedance cardiography and the rebreathing method during exercise in man. Jpn. J. Physiol. **39**(3), 441–446 (1989)
37. Nishimura, R.A., Callahan, M.J., Schaff, H.V., Ilstrup, D.M., Miller, B.A., Tajik, A.J.: Noninvasive measurement of cardiac output by continuous-wave Doppler echocardiography: initial experience and review of the literature. Mayo Clin. Proc. **59**, 484–489 (1984)
38. Rosenberg, P., Yancy, C.W.: Noninvasive assessment of hemodynamics: an emphasis on bioimpedance cardiography. Curr. Opin. Cardiol. **15**(3), 151–155 (2000)
39. Scherhag, A., Kaden, J.J., Kentschke, E., Sueselbeck, T., Borggrefe, M.: Comparison of impedance cardiography, thermodilution-derived measurements of stroke volume, cardiac output at rest, during exercise testing. Cardiovasc. Drugs Ther. **19**(2), 141–147 (2005)
40. Smorawiński, J., Nazar, K., Kaciuba-Uscilko, H., Kamińska, E., Cybulski, G., Kodrzycka, A., Bicz, B., Greenleaf, J.E.: Effects of 3-day bed rest on physiological responses to graded exercise in athletes and sedentary men. J. Appl. Physiol **91**, 249–257 (2001)
41. Willemsen, G.H., De Geus, E.J., Klaver, C.H., Van Doornen, L.J., Carroll, D.: Ambulatory monitoring of the impedance cardiogram. Psychophysiology **33**(2), 184–193 (1996)
42. Ziemba, A.W., Chwalbińska-Moneta, J., Kaciuba-Uscilko, H., Kruk, B., Krzeminski, K., Cybulski, G., Nazar, K.: Early effects of short-term aerobic training. Physiological responses to graded exercise. J. Sports Med. Phys. Fitness **43**(1), 57–63 (2003)

Chapter 5
Clinical and Physiological Applications of Impedance Cardiography Ambulatory Monitoring

The chapter contains the results of the clinical and physiological experiments which illustrate the possible fields of applications of the ambulatory impedance cardiography system.

5.1 Introduction

In my own studies ReoMonitor was used in the examinations performed in patients with different type of arrhythmia (atrial fibrillation, ventricular extrasystole beats), patients with implanted pacemaker and those who underwent the tilt-table testing. The results of the clinical and physiological experiments illustrate the possible fields of applications of the ambulatory ICG system. Also, the results obtained using other systems are described in this chapter.

The following five series of experiments (total 76 patients) were performed in which the cardiac hemodynamic parameters were ambulatory monitored using ReoMonitor [15, 19]:

1. the variability of stroke volume and ejection time were evaluated in 12 patients with atrial fibrillation (AF) and in 6 healthy male,
2. the hemodynamic effect of ventricular extrasystole beats was measured in 17 patients before anti-arrhythmic or ablative therapy,
3. the procedure of cardiac pacing optimisation (in a case study),
4. the hemodynamic effect of pacemaker syndrome was monitored in four patients who had the suspected diagnosis of pre-syncope, chest pain or dyspnoea,
5. the hemodynamic changes were monitored in 42 patients during the tilt test.

Additionally, some data obtained in 358 subjects of different age during static and dynamic exercise tests, psychological examinations and cold-pressor test were presented. In these experiments the ReoMonitor was used instead the stationary device, as a more comfortable device.

G. Cybulski, *Ambulatory Impedance Cardiography*,
Lecture Notes in Electrical Engineering, 76, DOI: 10.1007/978-3-642-11987-3_5,

5.2 Atrial Fibrillation

Atrial fibrillation is the most commonly encountered arrhythmia in clinical practice. In both, paroxysmal and persistent forms, AF leads to an increased rate of mortality [28, 32, 42, 46, 48, 75, 90]. There is an increasing awareness that AF is a major cause of embolic events, which in 75% of cases are complicated by cerebrovascular accidents [28, 42, 46].

AF is often associated with heart disease but a significant proportion of patients (about 30%) have no detectable heart disease [46, 79]. Symptoms, occasionally disabling hemodynamic impairment and a decrease in life expectancy, are among the untoward effects of AF, resulting in an important morbidity, mortality and an increased cost of health care [27, 55, 68].

Siebert et al. [75] studied the variability of SV as a response to the changes in body position in patients with coronary artery diseases and in healthy subjects. They analysed the power spectrum components of SVV (low-frequency band, high-frequency band and the ratio between them) in 60 patients before and at 6 weeks after CABG using the autoregressive method. They did not notice any significant changes in stroke volume spectral power indices before CABG. After CABG, all spectral indices were significantly decreased in the standing position.

It seems that the simple parameter of SV variability could be an indicator of the mechanical efficiency of the heart in patients with AF [50, 51]. Adding the ambulatory monitoring of the hemodynamics during AF might bring additional information describing the level of this impairment during the everyday life activity of the patient [2].

Verification of the ICG method in patients with AF was performed by Malmivuo et al. [56] who compared impedance and Fick methods in 11 patients with AF and without intracardiac shunts or valvular insufficiencies. They obtained the regression function $CO_Z = 1.05 \cdot CO_F + 0.1$, with a correlation coefficient of $r = 0.96$. Also Miyamoto et al. [58] checked his algorithm for automatic determination of hemodynamic parameters in patients with AF during spontaneous breathing at rest. Their comparisons between the computed stroke volumes and those obtained from a manual calculation "showed good agreement".

Pałko et al. [66] observed that CO in AF patients varied from 3.2 to 6.6 l min^{-1} (mean 5.3 ± 0.75 l min^{-1}) and increased by 12% after cardioversion (4.4–7.0 l min^{-1}, mean 5.9 ± 0.68 l min^{-1}).

The aim of the present study has been to describe the application of ambulatory monitoring system to determination of the variability of hemodynamic parameters during the different phases of the recording (daytime and night). The system made it possible to evaluate quantitatively the variability of the central hemodynamics for patients with AF and compare it with data obtained in healthy young volunteers. It was found that the coefficient of variation for SV and the amplitude of impedance signal were significantly higher in patients in comparison to controls' worst case. However, the coefficient of variation was not different for ejection time (ET) for all patients in comparison to controls. Thus it seems that the variability of

SV is mainly caused by the changes in amplitude of the signal (dz/dt_{max}) and that modification in ET has the smaller impact.

The simple parameter of SV (dz/dt_{max} and ET) variability reflects the level of SV (dz/dt_{max} and ET) modulation caused by AF and allows the quantitative evaluation of this phenomenon. It is especially important in the case of patients with paroxysmal AF. Through comparison of SV variability obtained during AF and the period of undisturbed heart activity it could be possible to evaluate the level of hemodynamic inefficiency caused by this type of arrhythmia. Also for patients with persistent AF the SV variability may be an indicator of the development of mechanical impairment or of the effectiveness of the applied therapy. Data obtained during normal activity of patients by means of the ICG holter method might be more useful in patients' diagnosis and therapy follow-up than those recorded during a short echocardiographic examination. Additionally, it is possible to evaluate the absolute decrease in SV caused by paroxysmal AF in comparison to the values obtained during sinus rhythm.

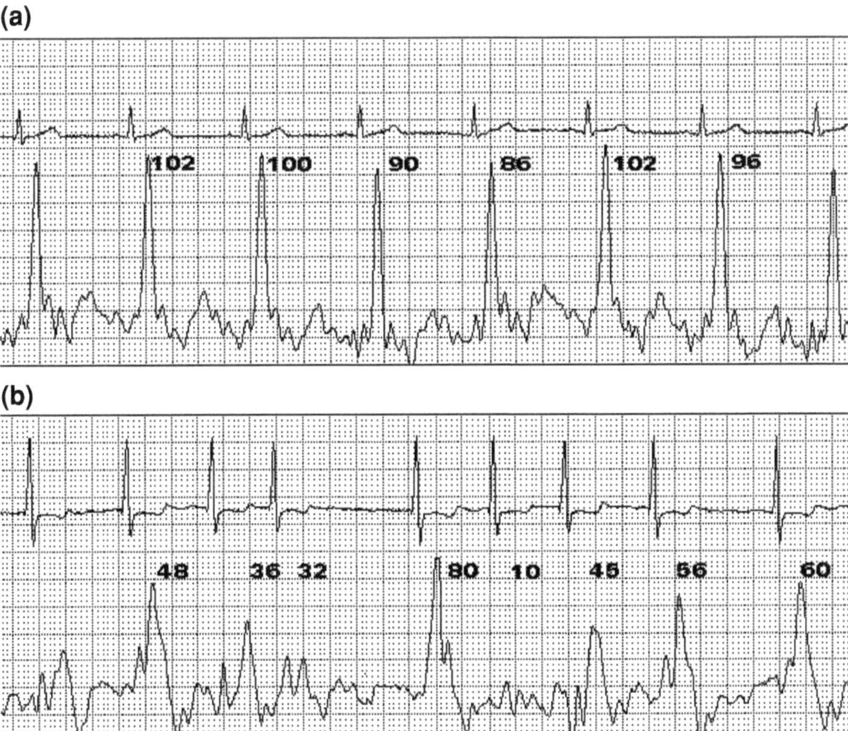

Fig. 5.1 ECG (first channel) and Impedance Cardiography dz/dt (second channel) traces (25 mm/s) recorded in normal healthy subject (**a**) and patient with atrial fibrillation (**b**). The highly variable amplitude of the dz/dt signal in patient is caused by the diminished venous return to the right atrium. Stroke volume expressed in (ml) is given besides the dz/dt curve for each cycle

Table 5.1 Beat-to-beat variations in hemodynamic parameters in a patient (66 years old) with atrial fibrillation. A recognised artefact is denoted as 0 in the last column (N/A)

Report

Time (s)	HR (1/min)	SV (ml)	CO (l/min)	RR (ms)	ET (ms)	PEP (ms)	$(dz/dt)_{max}$ (Ω/s)	Z0 (Ω)	N/A
00:54	89	76	6.8	670	245	165	2.39	20.1	1
00:55	82	48	3.9	725	230	185	1.62	20.1	1
00:56	114	8	0.9	525	100	999	0.63	20.1	0
00:56	120	17	2.0	500	100	999	1.34	20.1	0
00:57	92	70	6.4	650	240	165	2.25	20.1	1
00:58	77	47	3.6	775	235	175	1.56	20.1	1
00:58	82	33	2.7	725	220	200	1.17	20.1	1
00:59	78	52	4.1	765	250	175	1.63	20.1	1
01:00	86	38	3.3	690	230	180	1.29	20.1	1
Mean									
01:00	81	58	4.7	731	247	169	1.79	20.1	$N = 7$
SD	11	20	1.6	61	28	28	0.48	0	$A = 2$

In some earlier studies [18] and continued later [21] twelve male patients with AF aged 36–78 years and 6 healthy males (24–68 years) as a control underwent examination lasting for 12 h starting from about 1800 hours. Data were analysed from each 30 min period of recording during the daytime and over the night (0000–0400 hours). The variability was evaluated by calculating the distribution of the coefficient of variations of SV, dz/dt_{max}, and ET in each 30-min period of the night recording (4 h) and during the day (represented by one point).

Figure 5.1 presents ECG and ICG (dz/dt) traces recorded in a normal healthy subject (top strip) and in a patient with AF (bottom strip). The quantitative analysis showed that the highly variable amplitude of dz/dt signal in patient than in healthy person reflects the bigger fluctuations in SV. Moreover, the paroxysmal AF caused a 20–40% decrease in SV and similar changes in CO in comparison to the sinus rhythm in the same patient. An example of the printout of the report obtained for a 66-year-old patient with AF is presented in the Table 5.1.

Calculations performed on the both groups show the following: in patients it was lower SV (by 34 ml, $p < 0.005$, 87 ± 21 vs. 53 ± 18 ml), CO (1.79 l/min, $p < 0.001$, 5.40 ± 0.14 vs. 3.61 ± 1.6 l/min), amplitude of impedance signal dz/dt_{max} (0.65 ohm/s, $p < 0.01$, 1.77 ± 0.31 vs. 1.12 ± 0.40 ohm/s), and higher ET (28 ms, $p < 0.002$, 284 ± 52 vs. 312 ± 13 ms), PEP (47 ms, $p < 0.001$, 92 ± 21 vs. 138 ± 16 ms), and PEP/ET ratio (by 0.118, $p < 0.001$, 0.324 ± 12 vs. 0.442 ± 0.06).

The coefficient of variations (used as a index of variability) in patients was higher for SV (by 0.22, $p < 0.001$, 0.35 ± 01 vs. $0.55 \pm .27$), and for the amplitude of impedance signal dz/dt_{max} (0.17, $p < 0.005$, 0.26 ± 0.14 vs. 0.43 ± 0.22). The coefficient of variations was not different for ET (0.18 ± 0.05 vs. 0.19 ± 0.05).

Fig. 5.2 The distribution of the coefficient of variations describing the variability of stroke volume (SV) in each 30 min period during the night in AF patients (AF1–AF8) and in the worst case in healthy controls (N1–N4)

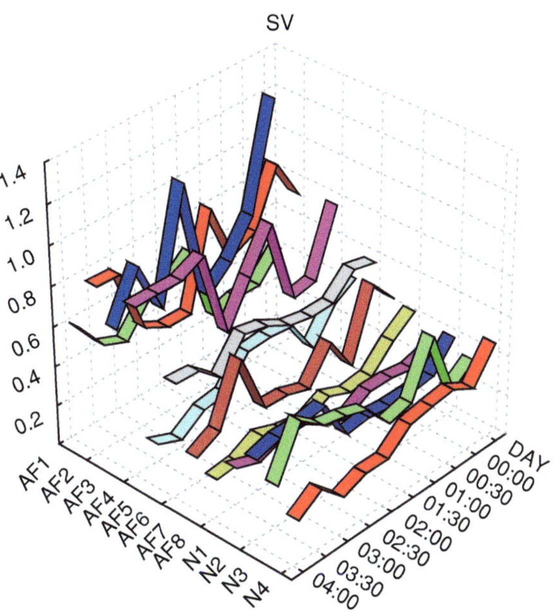

Fig. 5.3 The distribution of the coefficient of variations describing the variability of amplitude of impedance cardiography signals (dz/dt_{max}) in each 30 min period during the night in AF patients (AF1–AF8) and in the worst case in healthy controls (N1–N4)

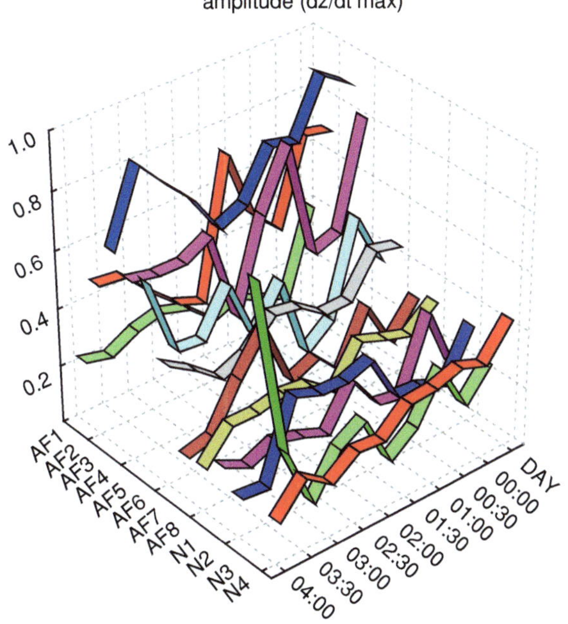

The distribution of coefficients of variation for SV, maximal amplitude of ICG signal, and ET in AF patients (AF1–AF8) and in the worst case in healthy control (N1–N4) are presented in Figs. 5.2, 5.3, 5.4 below. Data were obtained in each

Fig. 5.4 The distribution of
the coefficient of variations
describing the variability of
ejection time (ET) in each
30 min period during the
night in AF patients (AF1–
AF8) and in the worst case in
healthy controls (N1–N4)

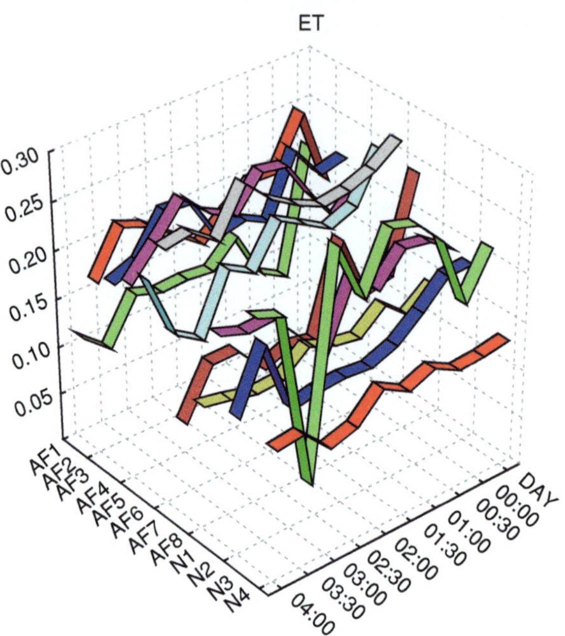

30 min period during the night between 00:00 and 04:00 a.m. Daytime recordings
are represented by a single point for signals obtained between 07:00–07:30 p.m.

From earlier [28, 42] and more recent [48, 79] studies it is known that stationary
(echocardiographically) detected reduced left ventricular systolic function is one
of the independent predictors of mortality in AF patients. This observation points
to the necessity of accurate quantitative measurement of left ventricular function
during AF episodes, particularly in its paroxysmal form.

It appears that the application of the central hemodynamics ambulatory monitoring system in patients with AF could give some additional diagnostic data
describing the level of impairment of cardiac mechanics caused by the paroxysmal
or persistent form of this arrhythmia.

5.3 Ventricular Extrasystole Beats (VEB) Monitoring

Patients with idiopathic ventricular extrasystolic beats (VEB) present various
clinical symptoms and variable arrhythmia behaviour at rest, during physical
activity and in daily life. Most frequently idiopathic ventricular arrhythmias
originate from the right ventricular outflow tract (RVOT) [3, 45]. RVOT
arrhythmias are usually well tolerated and associated with mild extrasystole pal-
pitations. There are no symptoms at all in up to 33% of patients [3]. Despite
exceptional reports of sudden death and ventricular fibrillation in adults and

children [31] these arrhythmias are treated as benign [4]. However, occasionally they can be very symptomatic and invalidating [70]. Takemoto et al. [84] suggested that frequent (>20%) RVOT-premature ventricular complexes may be a possible cause of LV dysfunction and/or heart failure. VEB may occur in a paroxysmal or a more regular form. Arrhythmia episodes with less effective VEBs may significantly decrease CO in comparison to sinus rhythm (SR).

If VEB is triggered by an unknown factor occurring during everyday activity it could be detected using an ECG Holter. However, with only ECG traces it is not possible to distinguish between hemodynamically efficient and non-efficient VEBs. Brockenbrough et al. [9] reported that the beat following premature ventricular contraction shows decreased pulse compared to the sinus rhythm. This is known as the Brockenbrough–Braunwald–Morrow sign or Braunwald sign [12]. However, there are only a few papers that describe quantitatively the level of CO decrease, with all its consequences for the patient. Sun et al. [80] used echocardiography to measure ejection fraction and cardiac index for normal beats and VEB in asymptomatic children without structural heart disease. They noted that in children with isolated monomorphic VEBs CO is markedly reduced if VEBs are frequent (>10/ min), have a short coupling interval or a prolonged QT interval. Stec et al. [78], in their case report measured CO using a pulsed Doppler technique in sinus rhythm $(4.4 \, 1 \, min^{-1})$ and during ventricular bigeminy $(2.9 \, 1 \, min^{-1})$. They reported that during bigeminy VEBs generated impact of $0.45 \, 1 \, min^{-1}$ whereas sinus beats in that period provided $2.45 \, 1 \, min^{-1}$. CO during ventricular bigeminy was 33% lower than in sinus rhythm. Satish et al. [70] in a case report wrote that intraortic pressure trace during ventricular bigeminy showed that VEBs "produced no detectable pressure".

However, some patients very rarely produce VEBs during echocardiographic examination, which makes quantitative evaluation impossible to perform. Ambulatory impedance cardiography allows continuous non-invasive evaluation of hemodynamic variables on a beat-to-beat basis during all daytime activity. There are, however, limited data on the usefulness of this method for the assessment of central hemodynamics in patients with idiopathic VEB during moderate exercise as a simulation of non-clinical conditions.

The purpose of our studies has been to evaluate the accuracy of ICG in the measuring hemodynamic parameters by comparing it with Doppler echocardiography and to assess which hemodynamical alterations can be responsible for arrhythmia-related symptoms [21, 22]. Seventeen adult patients were analysed during in-hospital evaluation before selection for antiarrhythmic or ablative therapy. A miniaturized, portable ICG device with a built-in one channel ECG (ReoMonitor) was used as a detector of central hemodynamic signals. Heart rate (HR), stroke volume (SV) and pre-ejection period (PEP) were obtained simultaneously by ICG and echocardiography in the supine position. Measurements were made in normal sinus beats and single, bigeminal or trigeminal VEB. Moreover, prolonged monitoring of ambulatory ICG was performed to analyse hemodynamical disturbance during symptoms and/or after 6-min walking test.

The discrepancies between PEP values obtained using the two methods were caused by the stiff rules of the automatic program for detection of both the

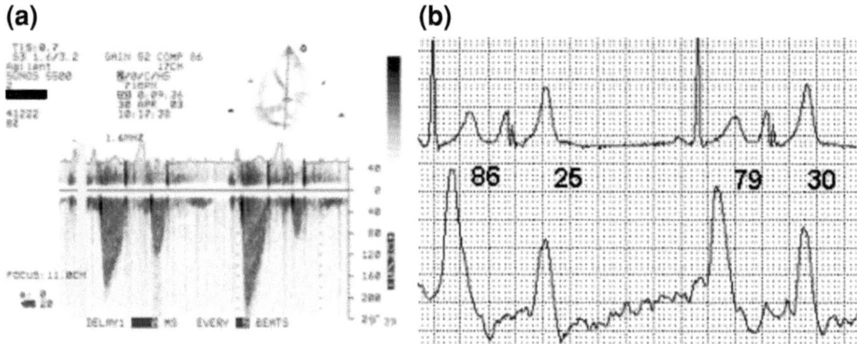

Fig. 5.5 The hemodynamic variability in bigeminy recorded using the pulsed Doppler method (*left*) and, in the same patient, ECG and ICG traces (*right*) (Adapted from [22])

beginning (Q-wave in extrasystole) and the end (opening the aortic valve) of PEP in VEB. Thus, PEP assessment requires further correction to produce a more efficient algorithm for automatic VEB evaluation. Precise calculations of PEP in VEB are still possible in manual mode. Hemodynamic monitoring obtained by portable ICG device may be useful for evaluation and management of asymptomatic and highly symptomatic patients with idiopathic VEB [77]. Ambulatory ICG recordings may also be used to distinguish between hemodynamically effective and ineffective extrasystole beats observed in the same patient during cardiac arrhythmia events (Fig. 5.7). It allows visualisation of the hemodynamic effect of paroxysmal arrhythmia (e.g. ventricular trigeminy, Fig. 5.6), which would be difficult to evaluate using gold standard, clinical monitoring methods. The hemodynamic consequences of this type of arrhythmia could be quantitatively evaluated and monitored using ambulatory impedance cardiography. Since Takemoto et al. [84] suggested that frequent (>20%) RVOT-premature ventricular complexes may be a possible cause of LV dysfunction and/or heart failure amenable to ablation therapy, ambulatory monitoring of CO in VEB could become an important diagnostic tool.

Measurement of SV by ambulatory ICG and Doppler were significantly correlated in sinus beats and VEBs ($r = 0.93$, $p < 0.001$ and $r = 0.74$, $p < 0.015$, respectively). Measurements of PEP were not correlated between ICG and Doppler in VEBs. Highly symptomatic patients at rest had a significant decrease in cardiac output during bigeminy revealed by both Doppler and ICG measurements. ICG allowed evaluation of various hemodynamic characteristics of simple VEBs and complex arrhythmia (interpolated VEBs, bigeminy, trigeminy, couplets) during daily life, while standing and after exercise. Figure 5.5 presents the hemodynamic variability in bigeminy recorded using the pulsed Doppler method and Ambulatory ICG in the same patient. Figure 5.6 presents the ECG and ICG traces acquired during ventricular trigeminy. This type of episode results in a 25–33% decrease in CO. Figure 5.7 illustrates the application of Ambulatory ICG to distinguish

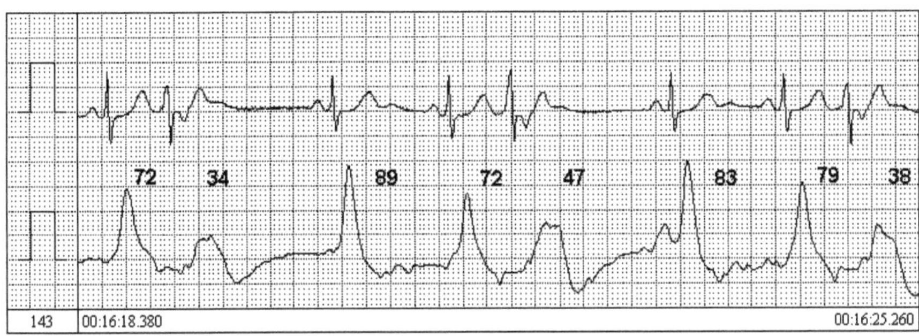

Fig. 5.6 The simultaneous recordings of one ECG channel and the impedance cardiography first derivative signal (dz/dt) during ventricular trigeminy. The numbers at the impedance trace denote stroke volume expressed in ml. This type of episode results in a 25–33% decrease in CO (Adapted from [22])

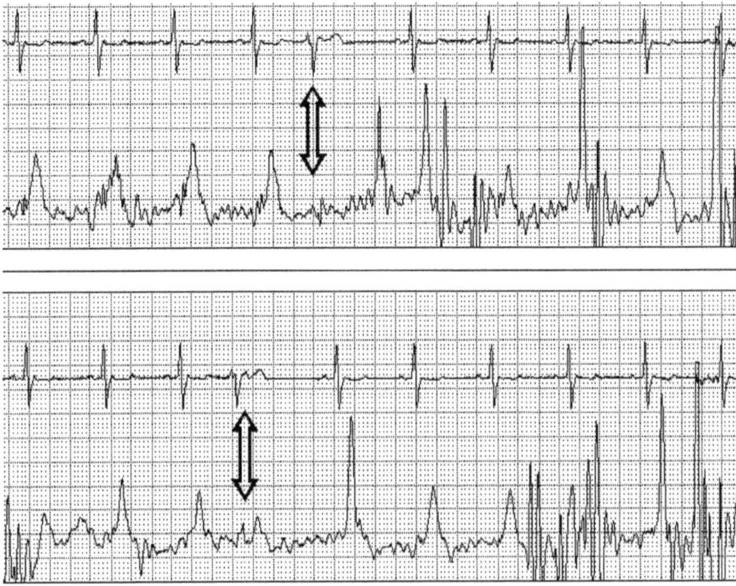

Fig. 5.7 Hemodynamically effective and ineffective extrasystole beats observed in the same patient during cardiac arrhythmia events (Adapted from [22])

between the hemodynamically effective and ineffective extrasystole beats observed in the same patient during cardiac arrhythmia events.

In patients with idiopathic ventricular extrasystoles ReoMonitor confirmed different pattern of SV and CO associated with symptomatic and asymptomatic bigeminy, single ventricular beats and ventricular interpolated and non-interpolated beats. Measurement of hemodynamic disturbances of ventricular ectopy was highly correlated with ECHO recordings in a supine position. Thus, a portable ICG device could be considered as a clinically acceptable and reproducible non-invasive method for assessment of SV and cardiac output in patients with idiopathic VEB.

5.4 Ambulatory ICG and Pacemaker Monitoring

5.4.1 Cardiac Pacing Optimisation

Dual chamber cardiac pacing (DDD) is increasingly used in the management of congestive heart failure. However, fixed differential atrioventricular delay (AVD), as offered by some manufacturers, are far from being physiological [35]. It was found that adjustment of AVD with respect to diastolic filling improves systolic function and is superior to fixed AVD settings. However, it is difficult to determine the optimal AVD in a particular patient with atrial and ventricular pacing. Moreover, inappropriate programming of the AVD decreases cardiac output significantly [62, 65, 86]. So, AVD optimisation is essential to maximise the hemodynamic benefits of this type of therapy. It could be performed using a non-invasive method of stroke volume determination.

In patients with obstructive hypertrophic cardiomyopathy AVD optimisation is used to decrease the transaortic gradient by changing the manner of chamber activation. Also advantages of AVD optimisation were observed in patients with outflow tract obstruction [87] though the mechanism of this phenomenon is not recognised. The long-term benefits of biventricular pacing have been observed [10, 11, 13, 36, 52, 67, 87] the decreased rate of mortality is not proven, but the standard of living improves and number of hospitalisations is diminished [7 62, 65]. AVD and interatrial hemodynamic optimisation is used in patients with atrial and ventricular resynchronisation pacing [GRAS et al. 1998; 24, 47, 82]. The hemodynamic effect of the Bachman bundle has also been analysed [1, 81]. In patients with AF hemodynamics are monitored to verify the effect of rhythm control, especially after ablation of the atrio-ventricular junction node [33, 83, 88]. Initial studies in patients with vasovagal syndrome showed that an ambulatory ICG method is useful in selection of the right program of cardiac pacing [39].

Figure 5.8 illustrates the case of a 64-year patient with a dual-chamber pacemaker monitored by ambulatory ICG during 12 h in the hospital environment [40]. Every 2 h the AVD was modified within the range of 160–300 ms. SV and CO were analysed during sinus rhythm (SR), DDD stimulation at $HR = 70$ beats^{-1}, and intervention $HR = 100$ beats^{-1}. On the basis of the ICG results the optimal AVD was set to 170 ms. The process of selection of the optimal atrio-ventricular delay (AVD) is presented in Fig. 5.8. Delays that are too short (AVD = 140 ms) or too long (AVD = 200 ms) result in decreased stroke volume (given in ml). The optimal AVD = 170 ms (middle strip) was selected when stroke volume was maximal.

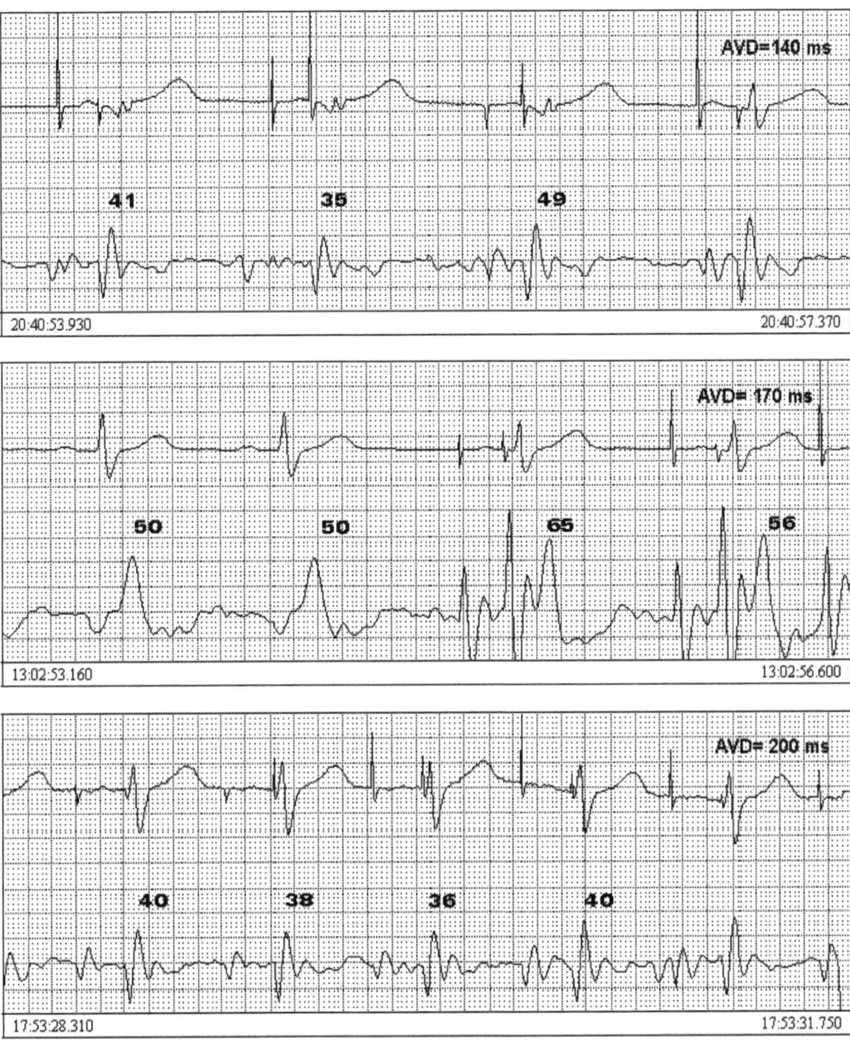

Fig. 5.8 Optimisation of atrio-ventricular delay (AVD) in patients with dual-chamber pacemakers using ReoMonitor. Excessively short (*top*) and excessively long (*bottom*) AVD result in decreased stroke volume (given in ml). The optimal AVD = 170 ms (*middle strip*) was selected when stroke volume was maximal (Adapted from [21])

Although Doppler echocardiography (echo) is the most widely used method to verify the pacemaker settings, it is time-consuming, expensive, and operator-dependent. [86]. ICG allows for non-invasive, continuous, beat-to-beat, quick, repetitive and operator independent evaluation of SV changes. Comparative studies performed during LV pacing using ICG and other methods of CO (or SV) evaluation showed that those measurements were closely correlated [25, 35, 38, 62–65, 86]. Moreover, in some medical centres stationary ICG is used to optimise

AVD not as a research but already as a routine method [65]. On the basis of the literature review and the author's own experience, it could be concluded that ICG:

1. allows highly reproducible non-invasive assessments of cardiac output in a pacemaker patient,
2. is a reliable method of AV interval optimisation during LV pacing and facilitates this procedure,
3. is well suited for routine examination of patients with cardiac dual chamber pacemakers.

Thus, an ambulatory version of ICG not only allows selection of the right AVD but may also confirm such a setting thanks to the availability of data obtained during normal activity of the patient. It could be used in all applications listed above when cardiac hemodynamics are essential for selection of the proper cardiac pacing program.

The application of pacemakers provides relief from life-threatening conduction disorders and arrhythmias as well as significantly improving the quality of life. However, they can function in a non-physiologic manner, which causes significant morbidity [60]. In some cases, the atrial contribution to the ventricular output is reduced, since the atrial contraction occurs against closed (AV) valves, producing reverse blood flow.

The stationary impedance cardiography is used for pacemaker adjustment in cardiac resynchronisation therapy (CRT) [37, 71]. There are no contradictions to use the ambulatory version of that method which could allow to verify the settings during everyday activity of the patient.

5.4.2 Pacemaker Syndrome Detection

Pacemaker syndrome is defined as intolerance to ventricular-based (VVIR) pacing due to loss of atrio-ventricular (AV) synchrony, which is associated with retrograde conduction or a reduction in SBP greater than 20 mmHg during pacing [29, 53, 57]. Although a definition of pacemaker syndrome is still a matter of discussion [23], its clinical symptoms include reproducible congestive signs, weakness or syncope. They are "the consequences of AV dyssynchrony or sub-optimal AV synchrony, regardless of the pacing mode" [23]. Pacemaker syndrome is associated with a decreased SV and, for the most part, occurs during slower heart action in a non-clinical condition. Confirmation of diagnosis of pacemaker syndrome requires simultaneous appearance of signs of AV asynchrony in ECG accompanied by hemodynamic disturbances.

Lamas et al. [44] investigated whether dual-chamber pacing would provide better event-free survival and quality of life than single-chamber ventricular pacing in patients with sinus-node dysfunction. They reported that clinical pacemaker syndrome was the principal reason for crossover from ventricular to dual chamber pacing in 18.3% of patients assigned to ventricular pacing and in 48.9%

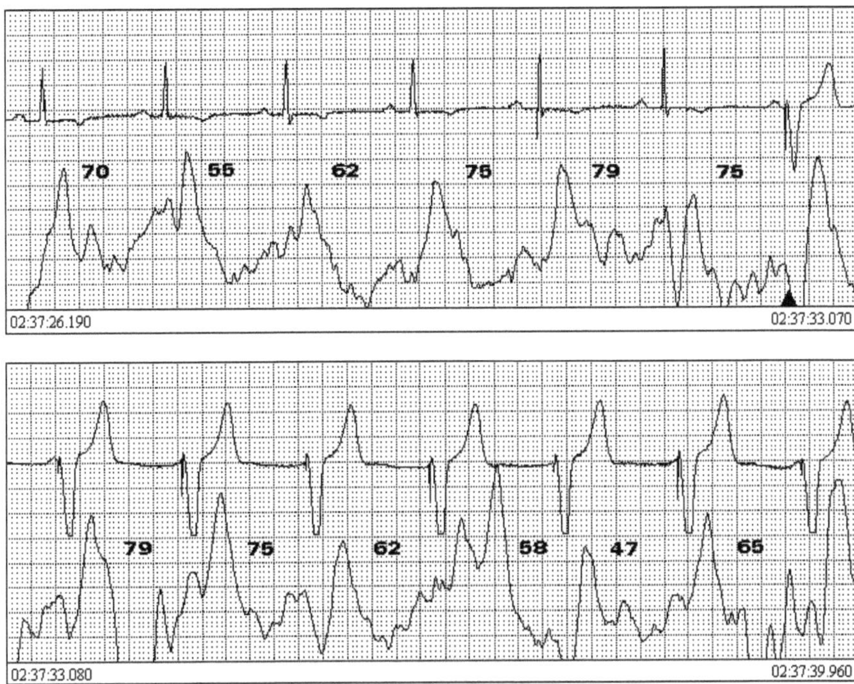

Fig. 5.9 The simultaneous recordings of one ECG channel and the impedance cardiography first derivative signal (dz/dt) in the patient with pacemaker. The top strip contains traces recorded during normal sinus rhythm (SR) and the bottom strip shows signals recorded when pacemaker syndrome occurred (third, fourth, and fifth cycle). The numbers at the impedance trace denote stroke volume expressed in ml (Adapted from [21])

of all patients who crossed over to DDD pacing. This may point out how important could be monitoring of cardiac hemodynamics in patients with ventricular pacing.

A hemodynamic ambulatory monitoring device could monitor changes in SV, which occur when a patient with pacemaker is outside a diagnostic lab. Moreover, an impedance cardiography holter, providing both ECG and mechanical activity signal, seems to be an excellent diagnostic tool that could help in distinguishing between pacemaker syndrome and other reasons for clinical symptoms (e.g. congestive or hypotension).

The ReoMonitor was applied to four patients with pacemakers who had the suspected diagnosis of pre-syncope, chest pain or dyspnoea. During long-term recordings the patients were asked to press a marker button to localise the occurrence of symptoms. In two patients diagnosis of pacemaker syndrome was made on the basis of simultaneous occurrence of hemodynamic disturbances correlated with clinical symptoms and ECG signs of atrio-ventricular asynchrony. Data from ReoMonitor were then verified by echocardiographic measurements.

Figure 5.9 contains the simultaneous recordings of one ECG channel and the impedance cardiography first derivative signal (dz/dt) in one pacemaker patient.

The top strip presents the traces recorded during normal sinus rhythm (SR) and the bottom strip shows signals recorded when pacemaker syndrome occurred. The numbers at the impedance trace denote stroke volume expressed in ml. The onset of stimulus triggering the QRS complex occurs just after atrial contraction. Thus, due to the loss of AV synchrony, the ineffective (for left ventricular filling) atrial contraction is performed against a closed AV valve which results in decreased SV. This may results in chest pain, weakness or syncope. During pacemaker syndrome SV is decreased.

ReoMonitor allowed confirmation of the hypothesis of pacemaker syndrome occurrence by simultaneous recording of hemodynamic decrease in SV and/or CO associated with atrio-ventricular asynchrony during decreasing sinus rhythm below 60 beats per minute when VVI pacing occurred. Parameters measured by Reo-Monitor were then confirmed by echocardiographic measurements.

5.5 Cardiac Parameters Monitoring During the Tilt Test

Orthostatic syndrome is a rapid and transient loss of consciousness accompanied by decrease in skeletal muscle tension. Although modern medical diagnostic equipment is used, still in 40% of patients the cause of orthostatic syndrome remains unrecognised [5, 8]. Application of a head-up tilt test (HUTT) allows selection of patients who have vasovagal syncope-impairment of the regulation processes in the cardiovascular system. The aetiology of this impairment is not fully recognized, which is caused by the lack of tools allowing continuous determination of parameters characterizing heart hemodynamics (stroke volume SV, arterial blood pressure BP). Moreover attempts to apply echocardiographic methods have failed, because they disturbed patients decreasing the test's sensitivity and long term monitoring is not possible due to errors caused by projection changes [5, 49, 54, 61, 91].

Some authors [8, 34] have pointed out the importance of HUTT in clinical practice. The mechanism of HR and BP changes and their relationship during the transient phase of the response to postural tests has been described in numerous papers [6, 26]. However the physiological mechanism of HR, SV and BP regulation and the pathophysiological causes of the vasovagal syncope remain unclear [8]. Also the length of the test and the peculiar circumstances of its application limit the number of patients that may be examined during 1 day.

Our tilt study was focused on looking for indices that allow fast and reliable diagnosis of vasovagal syncope. These could increase the number of tested patients per day and eliminate the monotony of the test. It appears that application of continuous monitoring of SV during HUTT using the impedance cardiography method in the ambulatory version (AICG) could help to find the relationship between the decrease of SV between in the initial period of the test and its final result.

Forty-two patients (26 female and 16 male, age 36 ± 16 years) underwent clinical tilt-table testing according to ESC Standards [8] because of unexplained syncope. Basing on the HUTT results patients were subdivided into two groups. The group with a positive test result (HUTT+) consisted of 15 patients (6 female, 9 male, age 35 ± 20) and the control group (HUTT−) of 27 patients (20 female, 7 male, age 36 ± 14) with no syncope during the tilt test. ECG and ICG signals were continuously recorded using ReoMonitor, which allows for beat-to-beat changes in HR and SV.

Absolute values of the HR, SV and CO at the last minute of resting did not reveal any differences between groups. Also, there were no significant differences in absolute values of HR, SV and CO, 1 min after the tilting. However in HUTT− group there was a non-significant decrease in SV below resting values 1 min after tilting. The greatest differences between HUTT− and HUTT+ were achieved in the fifth minute after tilting in SV. Changes in CO were caused mainly by changes in SV, because the HR response was similar in both groups. When absolute and percentage changes were considered the decline of SV and CO in HUTT+ was significantly more pronounced than in the HUTT− group (ΔSV: 27.2 ± 21.2 vs. 9.7 ± 27.2, $p = 0.03$; ΔCO: 1.78 ± 1.62 vs. 0.34 ± 2.48; $p = 0.032$; Δ%SV: 9 ± 51 vs. 34 ± 32; $p = 0.064$ (NS), Δ%CO: 30 ± 28 vs. 0.2 ± 58; $p = 0.034$) during the fifth minute after tilting.

Figure 5.10 presents beat-to-beat changes in heart rate (HR), stroke volume (SV), cardiac output (CO), ejection time (ET) and pre-ejection period (PEP) during a 60° tilt test in one subject. Arrows marks beginning and end of tilting (Table 5.2).

The continuous recording of the ECG and ICG traces enable to monitor the cardiac parameters during clinical tests. Figure 5.11 presents the ECG and ICG traces during 8 s asytolia observed in one of the patients during a 60° tilt test.

It was found that there were no differences between the HUTT+ and HUTT− groups in HR, SV, and CO at rest. Tilting provoked a different pattern of early hemodynamic response in these groups. In the HUTT− group there was a non-significant increase in SV and CO immediately after tilting, followed by a constant decrease in those parameters. In HUTT+ these parameters showed a tendency to decline in the whole early period of the response (5 min). In the fifth minute of the test a significant decrease in SV was observed for those patients who developed vasovagal syncope later during the test. This is in accordance with the finding of another study [30, 76] performed in young healthy men, who underwent lower body negative pressure (LBNP) stimulation. It was observed that those men who had low tolerance reacted earlier with markedly decreased SV. These findings and data from literature could suggest a relationship between the reaction in transient phase of the HUTT and vasovagal syncope. Novak et al. [61], using ICG, found a similar decrease of SV in two groups of patients (with and without positive HUTT outcome).

Shen et al. [74] found that baseline hemodynamic variables would not help in pre-selection of patients with vasovagal syncope, which is in accordance with the findings of the present study. However, before the symptom was developed during a tilt test a significant decrease of TPR in patients was observed whereas BP remained relatively stable. They suggested that distinct hemodynamic profiles in

Fig. 5.10 Beat-to-beat changes in heart rate (HR), stroke volume (SV), cardiac output (CO), ejection time (ET) and pre-ejection period (PEP) during a 60° tilt test on one subject. Arrows mark the beginning and the end of tilting (Adapted from [21])

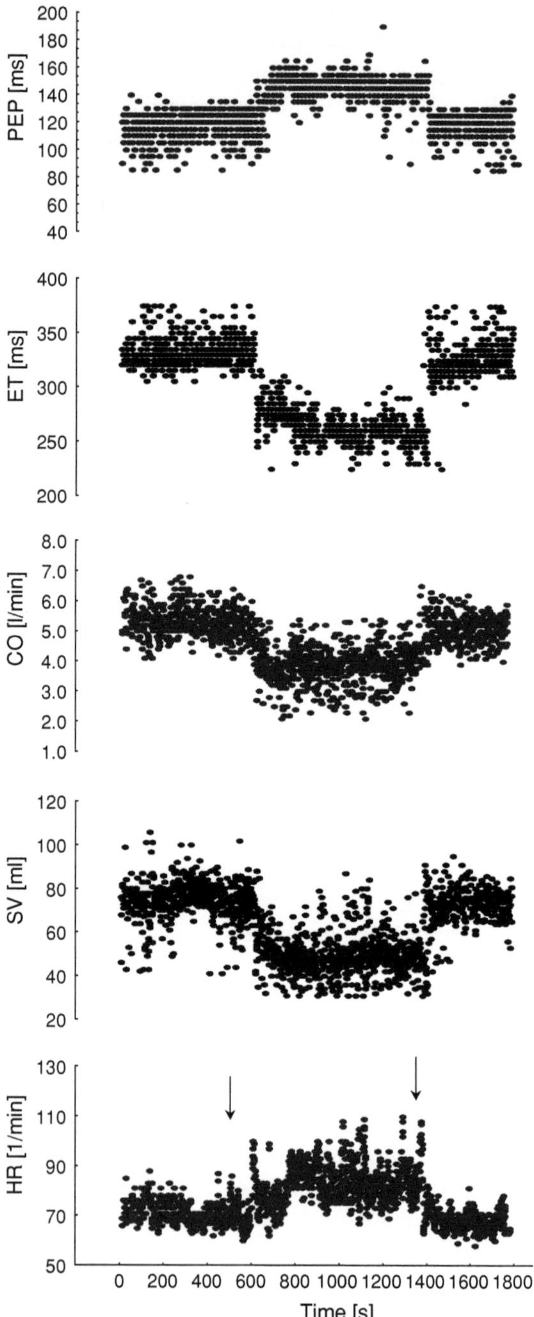

Table 5.2 The changes of the heart rate and hemodynamic parameters 5 min after tilting in comparison to values before tilting (adapted from [41])

	SV (ml)	CO ($1\,min^{-1}$)	ΔHR ($1\,min^{-1}$)	ΔSV (ml)	ΔCO ($1\,min^{-1}$)	Δ%HR (%)	Δ%SV (%)	Δ%CO (%)
HUTT−	53.6 ± 29.9	4.6 ± 2.82	6.4 ± 14.3	**9.7 ± 27.2**	**0.34 ± 2.48**	10 ± 17	9 ± 51	**0.2 ± 58**
HUTT+	41.8±18.6	3.36±1.44	7.7±13.4	**27.2±21.2**	**1.78 ± 1.62**	12±21	34±32	**30±28**
p	0.13	0.07	NS	**0.03**	**0.032**	NS	0.064	**0.034**

ΔHR increase of the heart rate 5 min after tilting, *ΔSV* decrease of stroke volume, *ΔCO* decrease of cardiac output, *Δ%HR* relative heart rate increase, *Δ%SV* relative stroke volume decrease, *Δ%CO* relative cardiac output decrease, *HUTT+* positive tilt-test, *HUTT−* control group without syncope during the tilt-test (statistically significant differences are marked using bold font)

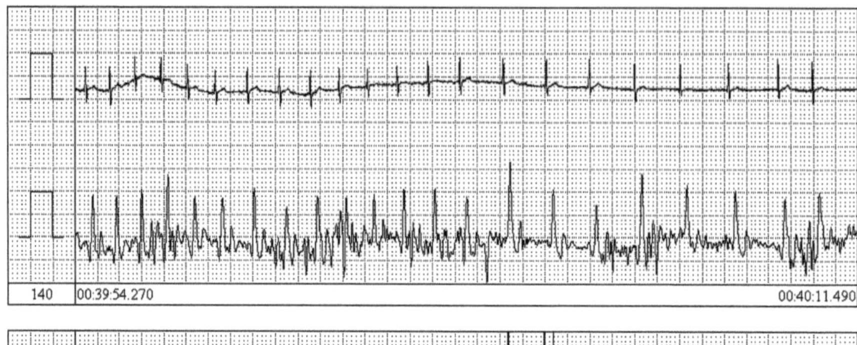

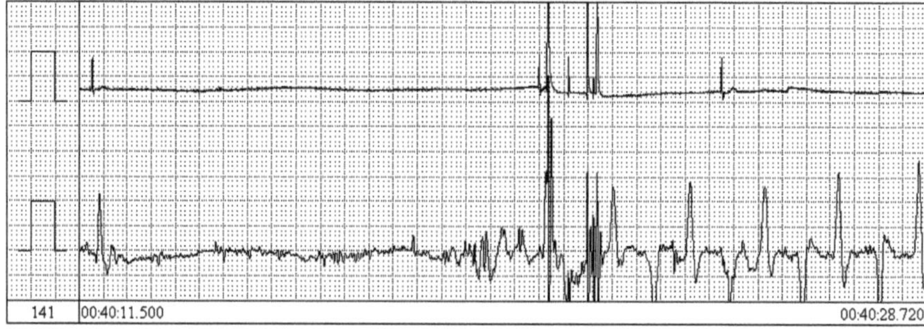

Fig. 5.11 ECG and ICG traces just before and during 8 s asytolia and observed in one of the patients during a 60° tilt test. The short recovery period is also visible. Please note the slowing down of the HR (speed 10 mm/s)

response to various provocative manoeuvres potentially could be defined by non-invasive, continuous monitoring.

Bellard et al. [5] used transthoracic impedance technique but did not calculate SV and CO. Instead, they used timing parameters (describing two phases of the ejection period), contractility index and an SV related parameter (the maximum amplitude of the first derivative of the impedance signal). There were differences between "fainters" and "non-fainters" at two points: in supine rest (for the slow phase of ejection) and just before the syncope (or at the end of the test in non-fainters) for the amplitude parameter. They did not observe any significant differences in the early phase of the HUTT between two groups. They also observed a tendency, similar in both groups, of decrease in the amplitude parameter (SV related) in the fifth-tenth minute of the test. In this study, SV also declined in the same time in both groups however the rate of this change was more pronounced in HUTT+. However, it is hard to find significant differences between these groups. The variance of hemodynamical parameters within the group is very high. However, the application of changes instead of absolute values enhanced the contrast between the groups. From the earlier experience in SV monitoring during postural stress [14, 20] it may be expected that ICG should also reveal differences in hemodynamic responses to tilt, between positive and negative HUTT patients, similar to those demonstrated in studies using echocardiography [49, 54, 91].

Considering the papers of [5, 74] and our findings, it could be concluded that ICG may demonstrate differences between the HUTT+ and HUTT− groups of patients. ICG technique seems well suited to monitoring relatively easily and continuously changes of stroke volume and some other cardiac hemodynamics parameters during potentially long lasting HUTT procedures [5, 14, 61, 74]. Moreover, the change in body position will not compromise the value of ICG measurement [20].

A practical effect of ICG application during a tilt test could be the shortening of the HUTT (at least potentially negative) from >60 min to 10–20 min and an increase in the number of the patients examined when cardiac hemodynamics (SV, CO) are monitored.

5.6 Other Applications

Ambulatory ICG has also been used in the author's home laboratory for assessment of the hemodynamic response to some physiological and clinical tests in healthy men and in patients. Krzemiński et al. [43], found differences in the

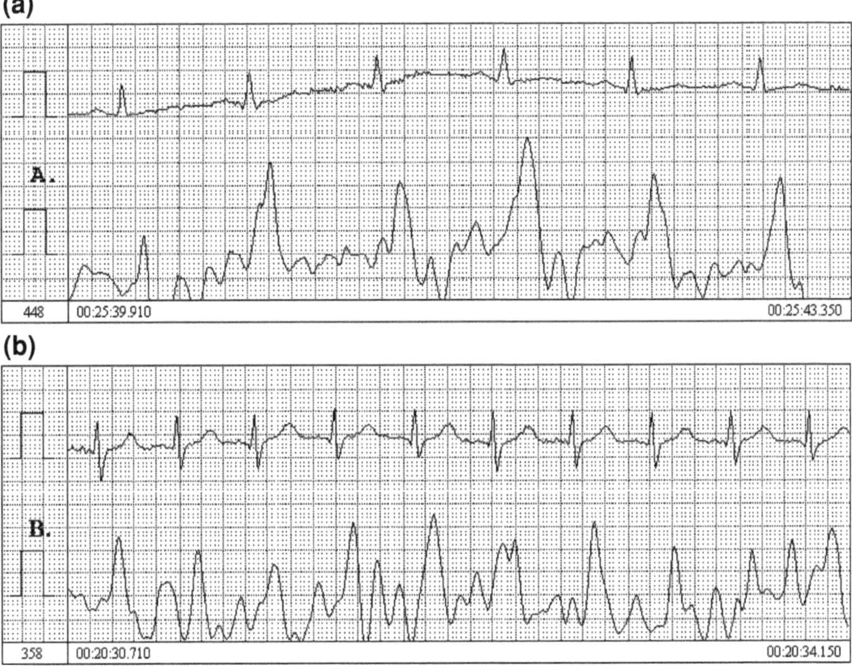

Fig. 5.12 The quality of the ECG and ICG traces during dynamic exercise on cycloergometer. The *upper strip* (**a**) was recorded in a 69 years old patient at the load of 80 W and the *bottom one* (**b**) in a 24 years old healthy man at the load of 150 W (Adapted from [21])

response of hemodynamic parameters to static exercise between eight male heart failure patients (functional class II/III NYHA, 67 ± 3.5 years) and eight healthy age-matched control male subjects (58 ± 7.4 years). Acceptable quality of recordings was obtained in both patients [59] and healthy subjects [76] during exercise tests on a cycloergometer (in a healthy man with intensity of up to 150 W). In the second study the system allowed identification of differences in response to the orthostatic manoeuvre and to dynamic exercise, after 3 days bed-rest. In another study, the system was used to follow the early hemodynamic effects of 6 weeks of endurance training [92].

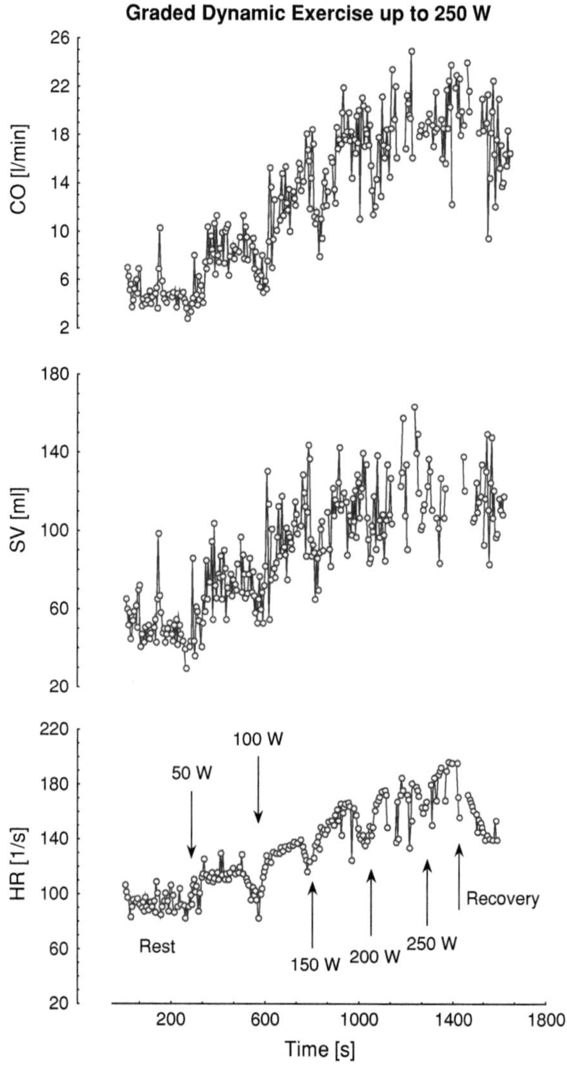

Fig. 5.13 The hemodynamic parameters recorded during graded dynamic exercise (up to 250 W) in a young healthy man (Adapted from [76])

The system was also applied in studies where hemodynamic response to psychological tests was monitored [59] and during a cold-pressor test in cardiac patients [16, 17].

Outside the home laboratory very promising data were obtained by Scherhag et al. [73], who found that hemodynamic measurements by ICG correlated highly significantly to simultaneous measurements by the thermodilution method at rest and during exercise testing. This study confirmed the usability of the ICG method for analysis of hemodynamic response to dynamic exercise. Some researchers use ambulatory versions of ICG to monitor the hemodynamic response to psychological load [89].

Figure 5.12 gives an example of the quality of ECG and ICG traces during dynamic exercise on a cycloergometer. The upper strip (A) was recorded in a 69 years old patient at a load of 80 W and the lower one (B) in a 24-year-old healthy man at a load of 150 W. The quality of the recordings obtained at 150 W in healthy man is still good, without a large number of artefacts and easy to process. These results indicate the usefulness of the monitoring of haemodynamical parameters during exercise testing on a cycloergometer, for example in patients undergoing cardiac rehabilitation programmes.

Figure 5.12 contains the hemodynamic parameters recorded during graded dynamic exercise (up to 250 W) in a young healthy man [76] (Fig. 5.13).

References

1. Bailin, S.J., Adler, S., Giudici, M.: Prevention of chronic atrial fibrillation by pacing in the region of Bachman's bundle: results of multicenter randomized trial. J. Cardiovasc. Electrophysiol. **12**, 912–17 (2001)
2. Barnes, V.A., Johnson, M.H., Treiber, F.: Temporal stability of twenty-four-hour ambulatory hemodynamic bioimpedance measures in African American adolescents. Blood Press. Monit. **9**(4), 173–177 (2004)
3. Belhassen, B., Viskin, S.: Idiopathic ventricular tachycardia and fibrillation. J. Cardiovasc. Electrophysiol. **4**, 356–358 (1993)
4. Belhassen, B.: Radiofrequency ablation of "benign" right ventricular outflow tract extrasystoles: a therapy that has found its disease? J. Am. Coll. Cardiol. **45**(8), 1266–8 (2005)
5. Bellard, E., Fortrat, J.O., Schang, D., Dupuis, J.M., Victor, J., Leftheriotis, G.: Changes in the transthoracic impedance signal predict the outcome of a 70 degrees head-up tilt test. Clin. Sci. (Lond) **104**, 119–126 (2003)
6. Borst, C., van Brederode, J.F.M., Wieling, W., van Montfrans, G.A., Dunning, A.J.: Mechanisms of initial blood pressure response to postural change. Clin. Sci. **67**, 321–327 (1984)
7. Braunschweig, F., Linde, C., Gadler, F., Rvden, L.: Reduction of hospital days by biventricular pacing. Eur. J. Heart Fail. **2**, 399–406 (2000)
8. Brignole, M., Alboni, P., Benditt, D., Bergfeldt, L., Blanc, J.J., Bloch Thomsen, P.E., van Dijk, J.G., Fitzpatrick, A., Hohnloser, S., Janousek, J., Kapoor, W., Kenny, R.A., Kulakowski, P., Moya, A., Raviele, A., Sutton, R., Theodorakis, G., Wieling, W.: Task Force on Syncope, European Society of Cardiology. Guidelines on management (diagnosis and treatment) of syncope. Eur. Heart J. **22**, 1256–1306 (2001)
9. Brockenbrough, E.C., Braunwald, E., Morrow, A.: A hemodynamic technic for the detection of hypertrophic subaortic stenosis. Circulation. **23**, 189–194 (1961)

10. Cazeau, S., Leclerq, Ch., Lavergne, T., et al.: Effects on multisite biventricular pacing in patients with heart failure and intraventricular conduction delay. NEJM **344**, 873–80 (2001)
11. Cazeau, S., Ritter, P., Bakdach, S., et al.: Four chamber pacing in dilated cardiomyopathy. PACE **17**, 1974–79 (1994)
12. Criley, J.M., Goldberg, S.L., French, W.J.: The Brockenbrough–Braunwald–Morrow sign. N. Engl. J. Med. **331**(23), 1589–1590 (1994)
13. Crisafulli, A., Melis, F., Lai, A.C., Orru, V., Lai, C., Concu, A.: Haemodynamics during a complete exercise induced atrioventricular block. Br. J. Sports Med. **36**(1), 69–70 (2002)
14. Cybulski, G.: Influence of age on the immediate cardiovascular response to the orthostatic maneouvre. Eur. J. Appl. Physiol. **73**, 563–572 (1996)
15. Cybulski, G., Książkiewicz, A., Łukasik, W., Niewiadomski, W., Pałko, T.: Central hemodynamics and ECG ambulatory monitoring device with signals recording on PCMCIA flash memory cards. Med. Biol. Eng. Comput. **34**, Suppl 1, part 1, 79–80 (1996)
16. Cybulski, G.H., Krysztofiak, J., Sawicki, W., Niewiadomski, M., Dłużniewski.: Impedance cardiography (holter) evaluation of the influence of phosphocreatine therapy on cardiac hemodynamics, III Conference of Noninvasive Cardiology Section of Polish Cardiac Society, Zakopane, 3–5 Apr 1997 Elektrofizjologia i Stymulacja Serca, **4** (1), 74 (in Polish) (1997)
17. Cybulski, G., Ziółkowska, E., Kodrzycka, A., Niewiadomski, W., Sikora, K., Książkiewicz, A., Łukasik, W., Pałko, T.: Application of impedance cardiography ambulatory monitoring system for analysis of central hemodynamics in healthy man and arrhythmia patients. Computers in Cardiology, IEEE, pp. 509–512. New York, NY, USA (1997)
18. Cybulski, G., Ziółkowska, E., Książkiewicz, A., Łukasik, W., Niewiadomski, W., Kodrzycka, A., Pałko, T.: Application of impedance cardiography ambulatory monitoring device for analysis of central hemodynamics variability in atrial fibrillation. Computers in Cardiology 1999. Vol. 26 (Cat. No.99CH37004), IEEE, pp. 563–566. Piscataway (1999)
19. Cybulski, G.: Ambulatory impedance cardiography: new possibilities, Letter to the Editor. J. Appl. Physiol. **88**, 1509–1510 (2000)
20. Cybulski, G., Michalak, E., Koźluk, E., Piątkowska, A., Niewiadomski, W.: Stroke volume and systolic time intervals: beat-to-beat comparison between echocardiography and ambulatory impedance cardiography in supine and tilted positions. Med. Biol. Eng. Comput. **42**, 707–711 (2004)
21. Cybulski, G.: Dynamic impedance cardiography—the system and its applications. Pol. J. Med. Phys. Eng. **11**(3), 127–209 (2005)
22. Cybulski, G., Stec, S., Zaborska, B., Niewiadomski, W., Gąsiorowska, A., Strasz, A., Kułakowski, P., Pałko, T.: Application of ambulatory impedance cardiography for analysis of ventricular extrasystole beats. In: WC 2009, IFMBE Proceedings 25/II, Fossel and Schlegel (Ed.). Springer, Verlag, pp. 712–714
23. Ellenbogen, K.A., Gilligan, D.M., Wood, M.A., Morillo, C., Barold, S.S.: The pacemaker syndrome—a matter of definition. Am. J. Cardiol. **79**, 1226–1229 (1997)
24. Etienne, Y., Mansourati, J., Touiza, A., et al.: Evaluation of left ventricular function and mitral regurgitation during left ventricular-based pacing in patients with heart failure. Eur. Heart Fail. **3**, 441–7 (2001)
25. Eugene, M., Lascault, G., Frank, R., Fontaine, G., Grosgogeat, Y., Teillac, A.: Assessment of the optimal atrio-ventricular delay in DDD paced patients by impedance plethysmography. Eur. Heart J. **10**(3), 250–5 (1989)
26. Ewing, D.J., Hume, L., Campbell, I.W., Murray, A., Neilson, J.M.M., Clarke, B.F.: Autonomic mechanisms in the initial heart rate response to standing. J. Appl. Physiol. **49**(5), 809–814 (1980)
27. Farshi, R., Kistner, D., Sarma, J.S.M., Longmate, J.A., Singh, B.N.: Ventricular rate control in chronic atrial fibrillation during daily activity and programmed exercise: a crossover open-label study of five drug regimens. JACC **33**, 304–310 (1999)
28. Feinberg, W.M., Blackshear, J.L., Laupacis, A., Kronmal, R., Hart, R.G.: Prevalence, age distribution and gender of patients with atrial fibrillation: analysis and implications. Arch. Intern. Med. **155**, 469–473 (1995)
29. Furman, S.: Pacemaker syndrome. Pacing Clin. Electrophysiol. **17**, 1–5 (1994)

30. Gasiorowska, A., Nazar, K., Mikulski, T., Cybulski, G., Niewiadomski, W., Smorawinski, J., Krzeminski, K., Porta, S., Kaciuba-Uscilko, H.: Hemodynamic and neuroendocrine predictors of lower body negative pressure (LBNP) intolerance in healthy young men. J. Physiol. Pharmacol. **56**(2), 179–193 (2005)
31. Haissaguerre, M., Shoda, M., Jais, P., et al.: Mapping and ablation of idiopathic ventricular fibrillation. Circulation **106**, 962–967 (2002)
32. Hardman, S.M., Noble, M.I., Biggs, T., Seed, W.A.: Evidence for an influence of mechanical restitution on beat-to-beat variations in haemodynamics during chronic atrial fibrillation in patients. Cardiovasc. Res. **38**(1), 82–90 (1998)
33. Ho, P.C., Tse, H.F., Lau, C.P., Hettrick, D.A., Mehra, R.: Effects of different atrioventricular intervals during dual-site right atria pacing on left atria mechanical function. PACE **23**, 1748–51 (2000)
34. Kenny, R.A., Ingram, A., Bayliss, J., Sutton, R.: Head-up tilt: a useful test for investigating unexplained syncope. Lancet **2**, 1352–54 (1986)
35. Kindermann, M., Frohlig, G., Doerr, T., Schieffer, H.: Optimizing the AV delay in DDD pacemaker patients with high degree AV block: mitral valve Doppler versus impedance cardiography. Pacing Clin Electrophysiol **20**(10 Pt 1), 2453–2462 (1997)
36. Kindermann, M., Frohlig, G., Doerr, T., Schieffer, H.: Optimizing the AV delay in DDD pacemaker patients with high degree AV block: mitral valve Doppler versus impedance cardiography. Pacing Clin. Electrophysiol. **20**(10 Pt 1), 2453–62 (1997)
37. Kułakowski, P., Makowska, E., Kryński, T., Stec, S., Czepiel, A., Błachnio, E., Soszyńska, M.: Tilt-induced changes in haemodynamic parameters in patients with cardiac resynchronisation therapy—a pilot study. Kardiol Pol. **67**(1), 19–24 (2009)
38. Kolb, H.-J., Böhm, U., Rother, T., Mende, M., Neugebauer, A., Pfeiffer, D.: Assessment of the optimal atrioventricular delay in patients withdual chamber pacemakers using impedance cardiography and Doppler. J. Clin. Basic Cardiol. **2**, 237–40 (1999)
39. Koźluk, E., Piątkowska, A., Wołkanin-Bartnik i wsp, J.: Stała stymulacja AAI u pacjentki z zespołem wazowagalnym—opis przypadku. Folia Cardiol. **8** streszczenia), 28 (2001)
40. Koźluk, E., Cybulski, G., Szufladowicz i wsp, E.: Zastosowanie Reomonitora w doborze optymalnego programu stymulatora z funkcją rate drop response. Folia Cardiol. **8**, streszczenia, 42 (2001)
41. Koźluk, E., Piątkowska, A., Kozłowski, D., Tokarczyk, M., Cybulski, G., Kamiński, R., Budrejko, S., Zapaśnik, P., Opolski, G.: Co wnosi monitorowanie parametrów hemodynamicznych i autonomicznego układu nerwowego w poznanie patofizjologii omdlenia wazowagalnego? Kardiol. Pol. **64**(8 Suppl. 5), S385–S386 (2006)
42. Krahn, A.D., Manfreda, J., Tate, R.B., Mathewson, F., Cuddy, T.E.: The natural history of atrial fibrillation: incidence, risk factors, and prognosis in the Manitoba follow-up study. Am. J. Med. **98**, 476–484 (1995)
43. Krzemiński, K., Kruk, B., Wójcik-Ziółkowska, E., Kozera, J., Cybulski, G., Nazar, K.: Effect of static handgrip on plasma adrenomedullin concentration in patients with heart failure, in healthy subjects. J. Physiol. Pharmacol. **53**(2), 199–210 (2002)
44. Lamas, G.A., Lee, K.L., Sweeney, M.O., Silverman, R., Leon, A., Yee, R., Marinchak, R.A., Flaker, G., Schron, E., Orav, E.J., Hellkamp, A.S., Greer, S., McAnulty, J., Ellenbogen, K., Ehlert, F., Freedman, R.A., Estes 3rd, N.A., Greenspon, A., Goldman, L.: Ventricular pacing or dual-chamber pacing for sinus-node dysfunction. N. Engl. J. Med. **346**(24), 1854–62 (2002)
45. Lauck, G., Burkhardt, D., Manz, M.: Radiofrequency catheter ablation of symptomatic ventricular ectopic beats originating in the right outflow tract. Pacing Clin. Electrophysiol. **22**(1 Pt 1), 5–16 (1999)
46. Lévy, S., Breithardt, R.W.F., Campbell, A.J., Camm, A.J., Daubert, J.-C., Allessi, M., Capucci, A., Cosio, F.: Atrial fibrillation: current knowledge and recommendations for menagement. Eur. Heart J. **19**, 1294–1320 (1998)
47. Leclercq, C., Cezeau, S., Ritter, P., et al.: Pilot experience with permanent biventricular pacing to treat advanced heart failure. Am. Heart J. **140**, 862–70 (2000)

48. Leonardi, M., Bissett, J.: Prevention of atrial fibrillation. Curr. Opin. Cardiol. **20**(5), 417–23 (2005)
49. Leonelli, F.M., Wang, K., Evans, J.M., Patwardhan, A.R., Ziegler, M.G., Natale, A., Kim, C.S., Rajikovich, K., Knapp, C.F.: False positive head-up tilt: hemodynamic and neurohumoral profile. JACC **35**(1), 188–193 (2000)
50. Leslie, S.J., McKee, S., Newby, D.E., Webb, D.J., Denvir, M.A.: Non-invasive measurement of cardiac output in patients with chronic heart failure. Blood Pressure Monitoring **9**(5), 277–280 (2004)
51. Lewis, R.P., Boudoulas, H., Leier, C.V., Unverferth, D.V., Weissler, A.M.: Usefulness of the systolic time intervals in cardiovascular clinical cardiology. Trans. Am. Clin. Climatol. Assoc. **93**, 108–20 (1981)
52. Linde, C., Leclerq, C., Rex, S., et al.: Long-term benefits of biventricular pacing in congestive heart failure: results from the Multisite stimulation in cardiomyopathy (MUSTIC) study. JACC. **40**, 111–8 (2002)
53. Link, M.S., Hellkamp, A.S., Estes, III N.A.M., Orav, E.J., Ellenbogen, K.A., Bassiema, I., Greenspon, A., Rizo-Patron, C., Goldman, L., Lee, K.L., Lamas, G.A.: High incidence of pacemaker syndrome in patients with sinus node dysfunction treated with ventricular-based pacing in the Mode Selection Trial (MOST). JACC **43**(11), 2066–2071 (2004)
54. Liu, J.E., Hahn, R.T., Stein, K.M., Markovitz, S.M., Okin, P.M., Devereux, R.B., Lerman, B.B.: Left ventricular geometry and function preceding neurally mediated syncope. Circulation **101**, 777–783 (2000)
55. Mackstaller, L.L., Alpert, J.S.: Atrial fibrillation: a review of mechanism, etiology and therapy. Clin. Cardiol. **20**, 640–650 (1997)
56. Malmivuo, J.A., Orko, R., Luomanmäki, K.: Validity of impedance cardiography in measuring cardiac output in patients with atrial fibrillation. In: Uusitalo, A., Saranummi, N. (eds.) Proceedings of The III Nordic Meeting On Medical and Biological Engineering, pp. 58.1–58.3. Finnish Society for Medical and Biological Engineering, Tampere (1975)
57. Mitsui, T., Hori, M., Suma, K., Wanibuchi, Y., Saigusa, M.: The "pacemaking syndrome." In: Jacobs, J.E. (ed.) Proceedings of the Eighth Annual International Conference on Medical and Biological Engineering, pp 29–33. Association for the Advancement of Medical Instrumentation, Chicago (1969)
58. Miyamoto, Y., Hiuguchi, J., Abe, Y., Hiura, T., Nakazono, Y., Mikami, T.: Dynamics of cardiac output and systoloic time intervals in supine and upright exercise. J. Appl. Physiol. **55**(6), 1674–1681 (1983)
59. Nazar, K., Kaciuba-Uściłko, H., Wójcik-Ziółkowska, E., Kruk, B., Pawłowska-Jenorowicz, W., Bijak, M., Kodrzycka, A., Krysztofiak, H., Niewiadomski, W., Cybulski. G.: Stress in work activity and health danger in patients with heart failure and the metabolic diseases (in Polish). Bezpieczeństwo Pracy, nr **3**, 6–9 (2001)
60. Newman, D.: Relationships between pacing mode and quality of life: evidence from randomized clinical trials. Card. Electrophysiol. Rev. **7**(4), 401–405 (2003)
61. Novak, V., Honos, G., Schondorf, R.: Is the heart "empty' at syncope? Auton. Nerv. Syst. **60**(1–2), 83–92 (1996)
62. Ovsyshcher, I., Gross, J.N., Blumberg, S., Andrews, C., Ritacco, R., Furman, S.: Variability of cardiac output as determined by impedance cardiography in pacemaker patients. Am. J. Cardiol. **72**(2), 183–7 (1993)
63. Ovsyshcher, I., Gross, J.N., Blumberg, S., Furman, S.: Precision of impedance cardiography measurements of cardiac output in pacemaker patients. Pacing Clin. Electrophysiol. **15**(11 Pt 2), 1923–6 (1992)
64. Ovsyshcher, I., Gross, J.N., Blumberg, S., Furman, S.: Orthostatic responses in patients with DDD pacemakers: signs of autonomic dysfunction. Pacing Clin. Electrophysiol. **5**(11 Pt 2), 1932–6 (1992)
65. Ovsyshcher, I., Zimlichman, R., Katz, A., Bondy, C., Furman, S.: Measurements of cardiac output by impedance cardiography in pacemaker patients at rest: effects of various atrioventricular delays. J. Am. Coll. Cardiol. **21**(3), 761–7 (1993)

66. Pałko, T., Kompiel, L., Ilmurzynska, K.: Stroke volume estimation by rheo-echocardiography during atrial fibrillation converted to sinus rhythm. Archiv. Acoust. **9**(1–2), 275–284 (1984)
67. Park, M.H., Gilligan, D.M., Bernardo, N.L., Topaz, O.: Symptomatic hypertrophic obstructive cardiomyopathy: the role of dual-chamber pacing. Angiology **50**, 87–94 (1999)
68. Pozzoli, M., Cioffi, G., Traversi, E., Pinna, G.D., Cobelli, F., Tavazzi, L.: Predictors of primary atrial fibrillation and concomitant clinical and hemodynamic changes in patients with chronic heart failure: A prospective study in 344 patients with baseline sinus rhythm. JACC **32**, 197–204 (1998)
69. Sakamoto, K., Muto, K., Kanai, H., Iiuzuka, M.: Problems of impedance cardiography. Med. Eng. Comput. **17**, 697–709 (1979)
70. Satish, O.S., Yeh, K.H., Wen, M.S., Wang, C.C.: Premature ventricular contraction-induced concealed mechanical bradycardia and dilated cardiomyopathy. J. Cardiovasc. Electrophysiol. **16**(1), 88–91 (2005)
71. Sciaraffia, E., Malmborg, H., Lönnerholm, S., Blomström, P., Blomström Lundqvist, C.: The use of impedance cardiography for optimizing the interventricular stimulation interval in cardiac resynchronization therapy—a comparison with left ventricular contractility. J. Interv. Card. Electrophysiol. **25**(3), 223–228 (2009)
72. Scherhag, A.W., Pfleger, S., de Mey, C., Schreckenberger, A.B., Staedt, U., Heene, D.L.: Continuous measurement of hemodynamic alterations during pharmacologic cardiovascular stress using automated impedance cardiography. J. Clin. Pharmacol. **37**(1 Suppl), 21S–28S (1997)
73. Scherhag, A., Kaden, J.J., Kentschke, E., Sueselbeck, T., Borggrefe, M.: Comparison of impedance cardiography and thermodilution-derived measurements of stroke volume and cardiac output at rest and during exercise testing. Cardiovasc. Drugs Ther. **19**(2), 141–7 (2005)
74. Shen, W.K., Low, P.A., Rea, R.F., Lohse, C.M., Hodge, D.O., Hammill, S.C.: Distinct hemodynamic profiles in patients with vasovagal syncope: aheterogeneous population. J. Am. Coll. Cardiol. **35**(6), 1470–7 (2000)
75. Siebert, J., Lewicki, L., Mlodnicki, M., Rogowski, J., Lango, R., Anisimowicz, L., Narkiewicz, M.: Atrial fibrillation after conventional and off-pump coronary artery bypass grafting: two opposite trends in timing of atrial fibrillation occurrence? Med. Sci. Monit. **9**(3), CR137–CR141 (2003)
76. Smorawiński, J., Nazar, K., Kaciuba-Uscilko, H., Kamińska, E., Cybulski, G., Kodrzycka, A., Bicz, B., Greenleaf, J.E.: Effects of 3-day bed rest on physiological responses to graded exercise in athletes and sedentary men. J. Appl. Physiol. **91**, 249–257 (2001)
77. Stec, S.B., Zaborska, G., Cybulski, P., Kułakowski, D., Piotrowska, K., Kuźnicka, W., Niewiadomski, E., Makowska.: Portable impedance cardiography monitor in patients with idiopathic ventricular extrasystoles. New clinical application and comparison with Doppler echocardiography. Pol. Heart J. **61**, III-254 (2004)
78. Stec, S., Walczak, F., Kulakowski, P., Zaborska, B., Rezler, J., Kokowicz, P.: Ventricular bigeminy originating from right ventricular outflow tract—treated with RF ablation (article in Polish). Kardiol Pol. **58**(6), 505–10 (2003)
79. Stollberger, C., Winkler-Dworak, M., Finsterer, J., Hartl, E., Chnupa, P.: Factors influencing mortality in atrial fibrillation. Post hoc analysis of an observational study in outpatients. Int. J. Cardiol. **103**(2), 140–144 (2005)
80. Sun, Y., Blom, N.A., Yu, Y., Ma, P., Wang, Y., Han, X., Swenne, C.A., van der Wall, E.E.: The influence of premature ventricular contractions on left ventricular function in asymptomatic children without structural heart disease: an echocardiographic evaluation. Int. J. Cardiovasc. Imag. **19**(4), 295–9 (2003)
81. Szili-Torok, T., Roelandt, J.R., Jordaens, L.J.: Bachman's bundle pacing: a role for three-dimensional intracardiac echocardiography. J. Cardiovasc. Electrophysiol. **13**, 97–8 (2002)
82. Szili-Torok, T., Bruining, N., Scholten, M., Kimman, G.J., Roelandt, J., Jordaens, L.: Effects of septal pacing on P wave characteristics: the value of three-dimensional echocardiography. PACE **26**, 253–6 (2003)

83. Takahashi, Y., Yoshito, I., Takahashi, A., et al.: AV nodal ablation and pacemaker implantation improves hemodynamic function in atria fibrillation. PACE **26**, 1212–7 (2003)
84. Takemoto, M., Yoshimura, H., Ohba, Y., Matsumoto, Y., Yamamoto, U., Mohri, M., Yamamoto, H., Origuchi, H.: Radiofrequency catheter ablation of premature ventricular complexes from right ventricular outflow tract improves left ventricular dilation and clinical status in patients without structural heart disease. J. Am. Coll. Cardiol. **45**(8), 1259–65 (2005)
85. Tsadok, S.: The historical evolution of bioimpedance. AACN Clin Issues. **10**, 371–84 (1999)
86. Tse, H.F., Yu, C., Park, E., Lau, C.P.: Impedance cardiography for atrioventricular interval optimization during permanent left ventricular pacing. Pacing Clin. Electrophysiol. **26**(1 Pt 2), 189–91 (2003)
87. Watanabe, K., Sekiya, M., Ikeda, S., Funada, J.I., Suzuki, J., Sueda, S., Tsuruoka, T.: Subacute and chronic effects of DDD pacing on left ventricular diastolic function in patients with non-obstructive hypertrophic cardiomyopathy. Jpn. Circ. J. **65**, 283–288 (2001)
88. Weerasooriya, R., Davis, M., Powell, A., et al.: The Australia Intervention Randomized Control of Rate In Atria Fibrillation Trial (AIRCRAFT). JACC **41**, 1697–702 (2003)
89. Willemsen, G.H., De Geus, E.J., Klaver, C.H., Van Doornen, L.J., Carroll, D.: Ambulatory monitoring of the impedance cardiogram. Psychophysiology **33**(2), 184–93 (1996)
90. Wolf, P.A., Mitchell, J.B., Baker, C.S., Kannel, W.B., D'Agostino, R.B.: Impact of atrial fibrillation on mortality, stroke and medical costs. Arch. Intern. Med. **158**, 229–234 (1998)
91. Yamanouchi, Y., Jaalouk, S., Shehadeh, A.A., Jaeger, F., Goren, H., Fouad-Tarazi, F.M.: Changes in left ventricular volume during head-up tilt in patients with vasovagal syncope: an echocardiographic study. Am. Heart J. **131**(1), 73–80 (1996)
92. Ziemba, A.W., Chwalbińska-Moneta, J., Kaciuba-Uscilko, H., Kruk, B., Krzeminski, K., Cybulski, G., Nazar, K.: Early effects of short-term aerobic training. Physiological responses to graded exercise. J. Sports Med. Phys. Fitness **43**(1), 57–63 (2003)

Chapter 6
Final Conclusions and Future Directions

In this chapter it is summarised the clinical importance of the ambulatory impedance cardiography monitoring and the prospects for the further development of the method.

6.1 Prospects for Impedance Ambulatory Monitoring

Oxygen saturation measured using pulse oximetry has emerged as a fifth vital sign (after ECG, BP, respiration, and temperature) that helps in determining the severity of a patient's condition. Now ICG is considered to be a sixth vital signal, one that could be used in Intensive Care Units. Similarly, widespread usage of an ICG signal in ambulatory monitoring as a method characterising cardiac mechanics in the natural environment of an active patient is only a question of time. The reason for this is not only the complex mechanical nature of e.g. paroxysmal arrhythmia [5], which could be followed in certain conditions but also the still not widely analysed morphology of the signal. For example, a small negative deflection from baseline, which occurs before aortic valve opening is strongly correlated with atrial contraction [8, 12, 13]. The maximum amplitude of the dz/dt signal reflects the peak aortic blood flow. So-called O-wave occurring in the diastolic part of the cardiac cycle is an indicator of venous return and cardiac filling [8, 12, 13]. And there is no other method, which could provide this kind of data using holter type of monitoring.

The process of changing impedance cardiography from laboratory applications to intensive care unit methods, from "bench to bedside" [11], started many years ago. A similar process will, perhaps, be observed for ambulatory version of this method, which appears to have numerous fields of applications: pharmacological dynamics monitoring [9], exercise physiology monitoring [1, 10], cardiac arrhythmia hemodynamics monitoring [1–3], etc.

G. Cybulski, *Ambulatory Impedance Cardiography*,
Lecture Notes in Electrical Engineering, 76, DOI: 10.1007/978-3-642-11987-3_6,
© Springer-Verlag Berlin Heidelberg 2011

When using an impedance cardiography ambulatory monitoring system, even the daytime recording, characterized by a markedly higher rate of artefacts in comparison to the night period observations, could provide a sufficient number of artefact-free cycles over a period of interest (e.g. arrhythmia events).

The presented data show that ambulatory impedance cardiography gives acceptable results in both absolute values of stroke volume and systolic time intervals (ET and PEP) measured during postural tests [4]. Availability of non-invasive ambulatory measures of cardiac function has the potential to improve care for a variety of patient populations, including those with hypertension, heart failure, pain, anxiety, and depressive symptoms [7]. Especially, very promising are the results that demonstrated that the AIM provides valid and reliable estimates of cardiac function in persons with hypertension [6].

6.2 Clinical Importance of the Ambulatory Impedance Cardiography Monitoring

Summarising the main achievements regarding the ambulatory impedance cardiography it may be concluded that:

1. A system for ambulatory monitoring of cardiac hemodynamics, consisting of a miniaturised, holter-type impedance cardiography device (with one built-in channel of ECG) and analysing software was successfully developed in several research centres. Some of the constructions were already commercialised and have the approval to use in clinical practice. Other could be used only in the research, non diagnostic field.
2. The system was verified using non-invasive (mainly ultrasound) and invasive methods. The ambulatory impedance cardiography system provides reliable measurement of hemodynamic parameters—cardiac output, stroke volume and systolic time intervals—in both supine and tilted positions. However some of the researchers suggest to limit the application of that method only to the measurement of the absolute values of systolic time intervals, and treat with reservations the absolute values of the stroke volume.
3. The system might be used to collect signals in a laboratory and in the field, for monitoring both the steady state and the transient phase of cardiovascular response to clinical and physiological tests using handgrip, cycloergometer, etc.
4. It was found that body movements and speech significantly distort ICG signal whereas recordings obtained during exercise performed on a cycloergometer are of good quality. This enables monitoring of hemodynamic parameters during exercise testing.
5. Application of the central hemodynamics ambulatory monitoring system in patients with atrial fibrillation and ventricular extrasystole beats, could give some additional diagnostic data describing the level of cardiac mechanics impairment caused by the paroxysmal or persistent form of these arrhythmias.

6. Ambulatory impedance cardiography may serve for optimisation of a–v delay in dual-chamber pacing systems at rest and for verification of the value of this parameter during normal daily activity. It could also help in distinguishing between pacemaker syndrome and other reasons for clinical symptoms.
7. Application of the ICG continuous hemodynamic monitoring during tilt testing may be helpful in diagnosing syncope and shortening test duration.
8. Ambulatory ICG might be a useful tool in evaluation the cardiovascular reaction to the stress tests to which the patient is exposed during a everyday activity (e.g. public speaking, arithmetic test, etc.).

Thus, ambulatory ICG may be helpful in the non-invasive assessment of hemodynamic impairment caused by cardiac arrhythmias and could serve for verification of VVI pacemakers and optimisation of a-v delay in dual-chamber pacing systems during normal daily activity. Ambulatory impedance cardiography could be used to record transient events, which would be difficult or even impossible to visualise using other, well established, "classical" methods. Moreover, it could be applied during the normal daily activity of the patient.

References

1. Crisafulli, A., Melis, F., Lai, A.C., Orru, V., Lai, C., Concu, A.: Haemodynamics during a complete exercise induced atrioventricular block. Br. J. Sports Med. **36**(1), 69–70 (2002)
2. Cybulski, G., Ziółkowska, E., Kodrzycka, A., Niewiadomski, W., Sikora, K., Książkiewicz, A., Łukasik, W., Pałko, T.: Application of impedance cardiography ambulatory monitoring system for analysis of central hemodynamics in healthy man and arrhythmia patients. In: Computers in cardiology, IEEE. 1997, pp. 509–512. New York, NY, USA (1997)
3. Cybulski, G., Ziółkowska, E., Książkiewicz, A., Łukasik, W., Niewiadomski, W., Kodrzycka, A., Pałko, T.: Application of impedance cardiography ambulatory monitoring device for analysis of central hemodynamics variability in atrial fibrillation. In: Computers in cardiology, vol. 26 (Cat. no. 99CH37004). IEEE. 1999, pp. 563–566. Piscataway, NJ, USA (1999)
4. Cybulski, G., Michalak, E., Koźluk, E., Piątkowska, A., Niewiadomski, W.: Stroke volume and systolic time intervals: beat-to-beat comparison between echocardiography and ambulatory impedance cardiography in supine and tilted positions. Med. Biol. Eng. Comput. **42**, 707–711 (2004)
5. Hardman, S.M., Noble, M.I., Biggs, T., Seed, W.A.: Evidence for an influence of mechanical restitution on beat-to-beat variations in haemodynamics during chronic atrial fibrillation in patients. Cardiovasc. Res. **38**(1), 82–90 (1998)
6. McFetridge-Durdle, J.A., Routledge, F.S., Parry, M.J., Dean, C.R., Tucker, B.: Ambulatory impedance cardiography in hypertension: a validation study. Eur. J. Cardiovasc. Nurs. **7**(3), 204–13 (2008)
7. Parry, M.J., McFetridge-Durdle, J.: Ambulatory impedance cardiography: a systematic review. Nurs. Res. **55**(4), 283–91 (2006)
8. Sramek, B.B.: Thoracic electrical bioimpedance: basic principles and physiologic relationship. Noninvas. Cardiol. **3**(2), 83–88 (1994)
9. Scherhag, A.W., Pfleger, S., de Mey, C., Schreckenberger, A.B., Staedt, U., Heene, D.L.: Continuous measurement of hemodynamic alterations during pharmacologic cardiovascular

stress using automated impedance cardiography. J. Clin. Pharmacol. **37**(1 Suppl), 21S–28S (1997)

10. Scherhag, A., Pfleger, S., Garbsch, E., Buss, J., Sueselbeck, T., Borggrefe, M.: Automated impedance cardiography for detecting ischemic left ventricular dysfunction during exercise testing. Kidney Blood Pressure Res. **28**, 77–84 (2005)

11. Summers, R.L., Shoemaker, W.C., Peacock, W.F., Ander, D.S., Coleman, T.G.: Bench to bedside: electrophysiologic and clinical principles of noninvasive hemodynamic monitoring using impedance cardiography. Acad. Emerg. Med. **10**(6), 669–80 (2003)

12. Tsadok, S.: The historical evolution of bioimpedance. AACN Clin. Issues **10**, 371–84 (1999)

13. Winter, U.J., Klocke, R.K., Kubicek, W.G., Niederlag, W. (eds.): Thoracic Impedance Measurements in Clinical Cardiology. Thieme Medical Publishers, New York (1994)

Appendix

Ambulatory Impedance Cardiography System: Reomonitor

Technical Specification of the Recorder

Analogue part	
The system of measurement	Alternating current, tetrapolar
Current generator	1 mA, 95 kHz, sinusoidal, stabilised, LC type
Output impedance	$R > 100$ kΩ
Number of the channels	4 (ECG, dz/dt, ΔZ, Z0)
Parameters of each channel	ECG (1 mV/1 V, f = 0–200 Hz, 3 dB, $R > 200$ MΩ)
	dz/dt (1 Ω/s/1 V, f = 0–15 Hz, 3 dB)
	ΔZ (100 mΩ/1 V, f = 0.2–40 Hz, 3 dB)
	Z0 (10 Ω/1 V)
Input impedance of the ICG receiver	>200 MΩ
Digital part	
Processor	80C552 Family micro-controller
	4×8 Bits A/D converters, input range 0–5 V
Sampling frequency	200 Hz
Memory medium	PCMCIA type II card, capacity >20 MB
	(Memory mapped mode)
Power consumption of digital part	10 mA during 99% of the recording time
Recording time	approx. 13 h for 20 MB card
Power source	9 V (6 $\times$ AA alkaline,
	or AA rechargeable batteries >1,200 mAh)
Physical specification	
Dimensions L $\times$ H $\times$ D (mm):	$200 \times 111 \times 50$
Weight (g)	665 g ($\sim$810 g with batteries),

Graphic Interface

Technical specification
  Programming environment   Windows 3.1 and upper
  2 modes of signal presentation   Bio-scope and Strip mode

Available options:

- independent on/off switching for each of the four channels,
- independent for each channel setting for zero level,
- four levels of amplification, according to the series: 0.5, 1.0, 1.5, 2.0,
- preliminary scaling factors according to the series: 0.125, 0.25, 0.5, 1.0, ..., 8.0,
- typical four levels of time scale: 10, 25, 50 and 100 mm/s plus full disclosure (FD-1-min per line),
- independent pre-setting of calibration parameters,
- static/scope presentation,
- colour of lines, background and grid selection,
- Two mode of selections of displayed strip: by scroll bar and by dialog window,
- cursor position display,
- calliper function for time and amplitude measurements,
- printing of a selected strip, fragment or full disclosure at different scaling levels: 33, 50, 75, 100, 150, 200 and 300%.

User-programmable features for data analysis module:
Display parameters (speed, amplification)
Separate grid lines on/off for amplitude and time calibration
Each channel separate pre-scaling: 0.125; 0.25;0.5; 1; 2; 4; 8.
Number of recording channels: up to 4
Data averaging time: 5–250 s
Data units: mV, Ω, Ω s^{-1}, mΩ (selection)

The Vrije Universiteit Ambulatory Monitoring System: Specifications and Features

This appendix was prepared based on the materials published by the Vrije Universiteit on their web page (http://www.psy.vu.nl/vu-ams/information/index.html)

The system allows to perform the following measurements:

- Heart rate measurement
- Beat-to-beat or averaged inter beat interval (IBI recording for any desired period

- Thorax impedance measurement to obtain respiratory signal and systolic time intervals
- Motility measurement as an indicator of the physical activity analysis.

Technical Specification

- Dimensions L × W × H: 120 × 65 × 32 mm.
- Weight with battery: 225 g
- Data memory: 2 MB solid state memory.
- Settings: wide range of sample rate settings for beat-to-beat recording, motility, GSR, thorax impedance, ICG ensembled average
- Recording time: 48 h for heart rate and motility at default sample rate settings.
- Current: 0.35 mA, 50 kHz
- Battery: 9-V Battery cell PP3, 14 mAh

Measurements

- Electrocardiography (ECG)

  - Full ECG is not recorded, RR intervals are stored with resolution of 1 ms
  - Sensitivity: 0.5–10 mV
  - Input impedance: 1 mV
  - Common mode rejection: >70 db
  - R–R time accuracy: 1 ms
  - R-wave detection procedure: auto trigger level

- Thorax impedance

  - Frequency: 50 kHz
  - Current: 350 µA
  - Range: 3–18 Ω
  - Resolution Z_0: 0.0233 Ω
  - Resolution ΔZ: 0.0030 Ω
  - Resolution dz/dt: 0.0010 Ω/s

- Motility (vertical acceleration)

  - Sensitivity: 3.2 gs (full scale)
  - Resolution: 0.008 gs
  - Frequency range: 1.6–50 Hz
  - Sample rate: 10–300 s

- Skin conductance level (SCL)

  - Range: 1–100 ms
  - Resolution: 0.0125 ms
  - Procedure: 0.5 V (constant voltage)
  - Sample rate:100–30,000 ms

Producer/Distributor: Vrije Universiteit Amsterdam

Division for Instrumentation of the Department of Psychophysiology,
Vrije Universiteit, FPP/ITM,
Van der Boechorststaat 1, room K1E-09,
1081 BT Amsterdam,
The Netherlands.
Web Page: http://www.psy.vu.nl/vu-ams/information/index.html
Phone: +31-20-5988854
Fax: +31-20-5988719
Email: vu-ams@psy.vu.nl

The MindWare MW1000A System: Specification and Features

All data given in this specification are taken from the materials published on the
web page http://mindwaretech.com

System Features

- The system collects waveforms for ECG, Z0, dZ/dt and GSC channel with 14
  bits resolution
- Computes and displays real time cardiac and systolic measures such as: LVET,
  PEP, SV, CO, HR, dZ/dt_{max}, Z0, mean inter beat interval (IBI) and RSA for hear
  rate variability, and skin conductance
- User-selectable sampling frequency: 500 or 1,000 Hz
- Source current specifications: 0.4 mA
- Precision oscillator: 100 kHz sinus waveform
- Built in 60 Hz hardware notch filter for ECG
- Real time waveform display on PDA
- Very low power consumption powered by internal PDA battery, optimizes run
  time
- User programmable collection epochs using simple tab delimited input file that
  can be edited on desktop and transferred to PDA
- The PDA ACQ software application can collect data in periodic or continuous
  modes
- User configurable settings for flexible acquisition
- File format compatible with MindWare desktop analysis applications IMP,
  HRV, and EDA.
- Personal data assistant: HP IPAQ

- Included accessories: patient cable, SD storage card (512 MB),
- Included software: data acquisition Wi-Fi Transmit software
- Physical dimensions: 45 × 95 × 160 mm
- Weight: 400 g
- Signal channel characteristics is presented in the Table A.1

Table A.1

Channel description	Channel symbol	Scale factor/gain	Units	Range
Impedance of thorax	Z0	0.05 V/Ω	Ω	0–40 Ω
First derivative of Z0	dZ/dt	0.8 V/s	Ω/s	0–3 Ω/s
Electrocardiogram	ECG	Gain = 500	V	0–2.5 V
Galvanic skin conductance	GSC	Gain = 125 μS/V	μS	0–200 μS

Mindware Technologies LTD
1020F Taylor Station Rd.
Gahanna, OH 43230, USA
Phone: +1-614-6264888
US Toll Free: 888 765-9735
Fax: +1-614-6264915
http://mindwaretech.com

The PhysioFlow Enduro System: Specification and Features

Size: 115 × 85 × 18 mm
Light weight: less than 200 g (with batteries)
Power supply: standard AA batteries or rechargeable, 6 h autonomy
Twenty-four-h MMC memory
USB or BlueTooth wireless memory download
Real time wireless monitoring using BlueTooth
The system works with PhysioFlow PF106 MS Windows[TM] XP based Software
for display, data analysis, and storage
Producer/ distributor:
Manatec Biomedical France
10bis, rue Jacob Courant
8300 Poissy
France
+33-9-65032401
http://www.physioflow.com/files/Fichiers_Manatec/Preliminary_brochure_Enduro.
pdf
NeuMeDx

2014 Ford Rd, Unit G
Bristol, PA 19007
USA
Tel.: +1-215-8269998
Fax +1-215-8268102
http://www.neumedx.com

AIM-8-V3 Wearable Cardiac Performance Monitor

This system is also known as "Ambulatory Bioelectric Impedance Monitoring for Assessing Cardiac Performance". This note was prepared based on their web page information and the scientific papers (Sherwood et al. 1998). According to the information from the producer the production of this type is discontinued although a several copies (April 2010) were still available.

Technical specification and features

Dimensions: 76 × 100 × 38 mm
Bioelectric impedance cardiograph (labeled the "WHIC8" for Wearable Hutcheson Impedance Cardiograph): tetrapolar, 80 kHz sine wave, 2 mA alternating current
   Microcomputer/data logger: credit card size
   Power supply: 9-V battery
   Power consumption: 200 μA
   From their web page (http://www.microtronics-bit.com/BIT/Products.html) we can learn that the AIM-8 device contains the firmware software that implements the following functions:

- Controls the various WHIC8 modes of operation, including calibration, operation on either a manual basis or an automatic fixed time-interval basis, and operation under a blood pressure monitor's control.
- Controls the displays, visual indicators, and audible alarms of the device.
- Detects user inputs and controls.
- Acquires and processes all analogue signals from the WHIC8 impedance cardiograph.
- Handles all timing for the system including real time, the timing of physiological waveforms and events (e.g. electrocardiogram R to R intervals), and data transfers.
- Performs all signal averaging and digital signal processing tasks.
- Handles the data analysis of the ensemble averaged waveforms and measured cardiac parameters.

- Performs the automatic computation of the various cardiac indices for each measurement.
- Manages the storage of ensemble averaged waveforms (Z_0, dZ/dt, and ECG), along with the "Data Scan" information regarding each cardiac cycle that makes up the ensemble averages.
- Stores all of the important measured parameters and the computed cardiac indices.
- Handles the serial data communication between the IBM-Type PC host computer and the AIM-8 monitor.

The AIM-8 system either measures or derives the following cardiac indices:

- Heart rate (HR)
- Stroke volume (SV)
- Cardiac output (CO)
- Pre-ejection period (PEP)
- Left ventricular ejection period (LVET)
- Cardiac contractility index (PEP/LVET)
- Basal impedance (Zo)
- Heather index (HI)
- Maximum amplitude of the first derivative of the impedance signal [$(dz/dt)_{max}$]
- Total peripheral resistance (TPR) (when blood pressure data are provided)
- Gross body activity-act (using body-activity sensor)

Producer/Distributor:
Bio-Impedance Technology,Inc.
88 VilCom Center, Suite 165
Chapel Hill, NC 27514
USA
http://www.microtronics-bit.com/BIT/Products.html

Ambulatory Impedance Cardiograph: AZCG

This information was provided basing on the papers by Nakonezny et al. (2001)

Technical Data and Specification

Dimensions: 45 × 95 x 160 mm
Weight: 400 g with batteries
Three-lead ECG: bandwidth 0.05–100 Hz
Four-lead electrical impedance system (tetrapolar):
Sine current of constant amplitude −2 mA, RMS at 50 kHz
Filters

Z_0: DC-100 Hz
ΔZ: DC-40 Hz
dz/dt: DC-40 Hz
The ECG and ICG have a digitally controlled, sampled-signal rebalance method for waveform stability.
The digital subsystem:
Motorola MC68332-based microcomputer
A–D conversion: 12-bit A–D converter, selectable sampling rate (100–1,000 Hz), 256 kB RAM
Data storage: 20 MB Flash Card (PCMCIA)
Ambulatory impedance cardiography AZCG
Producer/distributor: World Wide Medical Instruments

The Useful Links to Ambulatory Monitoring and Impedance Cardiography Web Pages

http://www.impedancecardiography.com
http://www.hemosapiens.com/teb.html
http://www.microtronics-bit.com/BIT/Products.html
http://mindwaretech.com
http://www.sonosite.com/products/bioz-dx
http://www.physioflow.com/clinicalinformation.htm
http://www.physioflow.com/files/Fichiers_Manatec/Preliminary_brochure_Enduro.pdf
http://www.neumedx.com/13.html
http://www.cardiolertsystems.com
http://www.psy.vu.nl/vu-ams/manuals/index.html (VU-AMS manual)
http://www.psy.vu.nl/vu-ams/information/index.html (VU-AMS web page)

Index